Textile Progress

2009   Vol 41   No 3

More about fibre friction and its measurements

Mehmet Emin Yuksekkaya

The Textile Institute

Taylor & Francis

Taylor & Francis

SUBSCRIPTION INFORMATION

Textile Progress (USPS Permit Number pending), Print ISSN 0040-5167, Online ISSN 1754-2278, Volume 41, 2009.

Textile Progress (www.tandf.co.uk/journals/TTPR) is a peer-reviewed journal published quarterly in March, June, September and December by Taylor & Francis, 4 Park Square, Milton Park, Abingdon, Oxon, OX14 4RN, UK on behalf of The Textile Institute.

Institutional Subscription Rate (print and online): $324/£170/€ 258
Institutional Subscription Rate (online-only): $308/£162/€ 245 (plus tax where applicable)
Personal Subscription Rate (print only): $119/£61/€ 95

Taylor & Francis has a flexible approach to subscriptions enabling us to match individual libraries' requirements. This journal is available via a traditional institutional subscription (either print with free online access, or online-only at a discount) or as part of the Engineering, Computing and Technology subject package or S&T full text package. For more information on our sales packages please visit www.tandf.co.uk/journals/pdf/salesmodelp.pdf.

All current institutional subscriptions include online access for any number of concurrent users across a local area network to the currently available backfile and articles posted online ahead of publication.

Subscriptions purchased at the personal rate are strictly for personal, non-commercial use only. The reselling of personal subscriptions is prohibited. Personal subscriptions must be purchased with a personal cheque or credit card. Proof of personal status may be requested.

Ordering Information: Please contact your local Customer Service Department to take out a subscription to the Journal: **India**: Universal Subscription Agency Pvt. Ltd, 101–102 Community Centre, Malviya Nagar Extn, Post Bag No. 8, Saket, New Delhi 110017. **USA, Canada and Mexico**: Taylor & Francis, 325 Chestnut Street, 8th Floor, Philadelphia, PA 19106, USA. Tel: +1 800 354 1420 or +1 215 625 8900; fax: +1 215 625 8914, email: customerservice@taylorandfrancis.com. **UK and all other territories**: T&F Customer Services, Informa Plc., Sheepen Place, Colchester, Essex, CO3 3LP, UK. Tel: +44 (0)20 7017 5544; fax: +44 (0)20 7017 5198, email: tf.enquiries@tfinforma.com.

Dollar rates apply to all subscribers outside Europe. Euro rates apply to all subscribers in Europe, except the UK and the Republic of Ireland where the pound sterling price applies. If you are unsure which rate applies to you please contact Customer Services in the UK. All subscriptions are payable in advance and all rates include postage. Journals are sent by air to the USA, Canada, Mexico, India, Japan and Australasia. Subscriptions are entered on an annual basis, i.e. January to December. Payment may be made by sterling cheque, dollar cheque, euro cheque, international money order, National Giro or credit cards (Amex, Visa and Mastercard).

Back Issues: Taylor & Francis retains a three year back issue stock of journals. Older volumes are held by our official stockists to whom all orders and enquiries should be addressed:
Periodicals Service Company, 11 Main Street, Germantown, NY 12526, USA. Tel: +1 518 537 4700; fax: +1 518 537 5899; email: psc@periodicals.com.

The 2009 US Institutional subscription price is $324. Periodical postage paid at Jamaica, NY and additional mailing offices. US Postmaster: Send address changes to TTPR, C/O Odyssey Press, Inc. PO Box 7307, Gonic NH 03839, Address Service Requested.

Subscription records are maintained at Taylor & Francis Group, 4 Park Square, Milton Park, Abingdon, OX14 4RN, United Kingdom.

For more information on Taylor & Francis' journal publishing programme, please visit our website: www.tandf.co.uk/journals.

CONTENTS

Textile Progress
Vol. 41, No. 3, 2009, 141–193

More about fibre friction and its measurements

Mehmet Emin Yuksekkaya*

Department of Textile Engineering, Usak University, Usak, Turkey

(Received 4 May 2009; final version received 10 July 2009)

Unfortunately, the classical empirical friction laws do not hold true for fibrous and viscoelastic materials comprising most of the textile fibres. In the second half of the twentieth century, fibre surfaces have been studied by many distinguished scientists who were able to complete numerous researches for the frictional characteristics of different types of fibres. Most of the researchers have aimed to develop a new test method and a test device that can be used to measure the frictional characteristics of fibres quickly, accurately and easily in their studies. Unfortunately, there is not a standard test method or a test device for the measurement of textile fibres' friction properties. For today's competitive marketing, the instrument for fibre testing must be very fast and accurate; otherwise, it will not be useful for commercial purposes. For example, hundreds of thousands of cotton bales should be tested within a very short period of time in terms of the length, colour and trash content of the cotton bales. Without having the data describing the properties of cotton fibres, cotton bales cannot be sold commercially in most of the countries. Therefore, it is an important factor that the fibre-testing instrument should be fast and accurate. Most of the properties of cotton fibres can be assessed by using a HVI fibre-testing instrument. In this review, the historical perspective of fibre friction studies has been demonstrated with the fibre friction measurement-testing devices.

Keywords: word; fibre friction; stick-slip; friction measurement

1. Introduction

Traditionally, most of the mechanical problems in elementary physics and mechanic courses were solved by ignoring the effect of friction because the inclusion of frictional forces made the analysis too complicated. Although friction has been ignored in many classic mechanical problems, it is a very important phenomenon in our daily life. Left to act alone, it is the frictional force that brings every rotating shaft to a halt. In today's automobiles, for instance, about 20% of the engine power is consumed to overcome internal frictional forces [1,2]. It seems that friction causes expenditure on unnecessary energy to overcome it. On the other hand, without friction, no one can walk, hold a pen or pencil and, even if they could, it would not allow writing. Furthermore, it would not be possible to have a wheeled transportation system without friction.

During the last three decades, the production speed of textile machinery has been improved. To make best use of this improvement, a precise measurement of the properties of fibres is required. The frictional characteristic of fibres is a very important property at all stages in the conversion of fibres into an end product. For example, in the spinning process, one metre of fibre mat is expanded approximately into 2000 metres of corresponding yarn.

*Email: meyuksek@sesli.com.tr

ISSN 0040-5167 print/ISSN 1754-2278 online
© 2009 The Textile Institute
DOI: 10.1080/00405160903178591
http://www.informaworld.com

During this enormously fast process, fibres interact with each other and with machine parts. With today's high-speed fibre-testing machines (i.e. High Volume Instrument (HVI) and Advance Fibre Information System (AFIS)), most of the fibre properties can be assessed easily, including their length, colour and thickness. The measurement of friction, on the other hand, is a challenging and critical issue not only for textile applications but also for many other high-tech industries. For example, it is necessary to know the frictional force very accurately for control systems [3], the metal forming industry [4–6], computer drivers [7], surgical tools, and civil engineering applications [8–10].

Measurement of the frictional forces has been a topic of interest to many workers for a long time. Among them, Leonardo da Vinci and Amontons are known as the first people to give a statement about their work related to the frictional phenomenon. However, it is quite possible that in the history of the world, there were some other scientists who had worked in this area, but either did not address friction, or their ideas may have been lost in the darkness of the history. Many distinguished scientists tried to explain the frictional phenomenon by introducing some relationship among the various physical quantities such as weight, size and surface characteristics of the object. The most famous relationship, $F = \mu N$, states that the frictional force F is proportional to the normal force N with a factor of μ that is a constant called the coefficient of friction [1,11,12]. In this fundamental equation, the surface characteristics of the objects have not been mentioned explicitly. This basic relation may approximately explain the characteristics of friction for most of plastic materials. However, this is not a complete explanation for friction; especially for fibrous or viscoelastic materials, this relationship does not hold under any circumstances.

Friction has an important role in textile applications. During the spinning process, for instance, it is the only force holding fibres together if no twist is present [13,14]. Yarns are constrained to move from process to process by specially designed guides. They are always interacting with tensioners in the winding and warping, passing through a heddle eye in weaving, needle eye in knitting and sewing etc. The interaction between yarns and machine parts can deteriorate the yarn quality and may result in machine stoppage later in the process. The magnitude of frictional properties of fibres has a critical importance in many textile processes. Depending on application type, it may be desired to have a high value of friction in some applications. In staple yarns, for instance, friction is the force that holds the fibres and gives strength to yarns. In fabrics, it is the force among the interlacing yarns [15–17]. It seems that a high value of friction force is necessary, but on the other hand, it may also introduce some operational difficulties during the spinning process if its value is high. Therefore, the frictional properties of fibres, namely, the frictional coefficients, must be low enough to prevent difficulties during processes and high enough to give cohesion among fibres. This phenomenon, known as 'Ball's Paradox' [18], was reported as follows: '*Up to the front mule roller, cotton must be slippery; afterwards it must be sticky*'.

As described above, depending upon the type of the application, friction is desirable or a nuisance. However, For spinning processes it must be in an acceptable range, that is, the coefficient of friction should be low to allow an easy process during drafting and high to give cohesion among the fibres so that fibres will not fall apart. This concept related to differences in 'static friction' and 'dynamic friction' was named as stick-slip phenomenon. The frictional forces should be expressed in terms of the properties of the body and its environment. Friction viewed at the microscopic level is a very complicated phenomenon. Force laws for sliding friction are empirical in character and approximate in their predictions. They do not have an elegant simplicity or accuracy that one can find for

Table 1.   Static and kinetic coefficient of friction for cotton fibres.

Investigator	Static coefficient (μ_s)	Kinetic coefficient (μ_k)	Difference $(\mu_s - \mu_k)$
Morrow	0.396	0.220	0.176
Mercer	0.570	0.316	0.254
Krowicki	0.452	0.249	0.203
McBride	0.54	0.300	0.240
Viswanathan	0.587	0.327	0.260
Bryant	0.523	0.293	0.230
Levy	0.500	0.275	0.225
Gunther	0.590	0.240	0.350
Belser	0.520	0.271	0.249
Belser	0.496	0.243	0.253
Cromer	0.490	0.250	0.240
Whitworth	0.453	0.250	0.198
Whitworth	0.386	0.173	0.213

the other physical phenomena, such as gravitational or electrostatic force laws. However, it is remarkable, considering the enormous diversity of surfaces one encounters, that many aspects of frictional behaviour can be understood qualitatively based on a few simple mechanisms.

In this text, the historical aspects of the measurement of the friction and the stick-slip phenomenon, together with the instrumentation used for those purposes, are given. Most of the friction models given in the literature are purely in empirical form. Therefore, to replicate the test done can be very difficult, and the data obtained from such a test may not be reproducible under given conditions [19–23]. This may be observed by examining the coefficient of the friction data given in Table 1.

2. Fundamental studies of friction

2.1. Introduction

The main interest of many workers led to the investigation of the actual physical characteristics of friction. In order to reduce the frictional forces acting on a metal surface, it is necessary to smooth the surface so that so-called surface adhesion does not occur. Figure 1 shows a highly magnified view of a section of a finely polished steel surface. As seen in Figure 1, even highly polished surfaces are far from plane when viewed on the atomic scale. The cross section of such a polished metal surface normally looks like a profile drawn as shown in Figure 1.

When two bodies are placed in contact, the actual microscopic area of the contact is much less than the apparent macroscopic area of the contact. The actual area of the contact is proportional to the normal force because the contact points deform plastically under heavy stress developed at these points. These contact points become 'cold-welded' and produce 'surface adhesion'. At the contact points, the molecules on the opposite sides of the surface are so close to each other that they exert strong intermolecular forces among the molecules. When one body is pulled across another, the frictional resistance is associated with the rupturing of these thousands of tiny welds that continually reform as new contacts

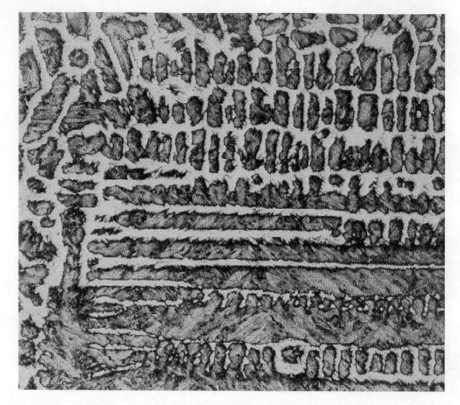

Figure 1. Highly magnified view of a section of a finely polished steel surface.

are made, as seen in Figure 2. It would be appropriate to divide the studies related to the friction into the following two groups:

- Friction of hard solids.
- Friction of fibrous and viscoelastic materials.

Textile fibres mostly fall into the second category in which the friction of fibrous and viscoelastic materials is valid.

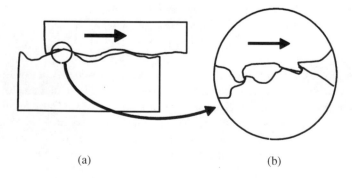

(a)   (b)

Figure 2. Details of sliding friction.

2.2. Principles of friction

A basic definition of friction given in physics books is '*the resistance encountered when two bodies are brought to contact and allowed to slide against each other*' [1,2,11,12,20,24–27]. In the history of science, two types of forces frequently encountered are gravitational and frictional forces. Gravitational force has been studied by great scientists in every age, whereas frictional force has been neglected [11]. It has been assumed that the sliding process holds little intrinsic interest and that the three simple laws adequately describe the force of friction. It might have been reasonable to neglect the frictional forces in the past. However, with the advent of modern machinery, working with very close tolerances under new and widely varying conditions has shown the inadequacy of the knowledge about the sliding process and hitherto neglected friction phenomenon. The three laws describing the force of friction for a solid body are as follows:

- The frictional force is proportional to the load, or pressure of the area of contact.
- It is independent of the area of contact.
- It is also independent of sliding speed.

The first two laws were stated by Leonardo da Vinci in the fifteenth century and reported by Guillaume Amontons in the 1690s. However, the third law, better known in electrostatics, was first given by Charles Augustine de Coulomb, the French physicist [1,2,11,12,20]. The first two laws generally hold true with no more than 10% deviation for metals when they are deformed plastically [11,28]. It has been known for some time that friction is dependent on sliding speed and it is claimed that the frictional force first increases with velocity and then falls, as seen in Figure 3 [12].

Hence, one interesting point relates to the mechanism between the coefficient of friction and time. In any adhesive process, the bond becomes stronger the longer it is left undisturbed. The static coefficient of friction increases with time of contact, which is an indication of the validity of adhesion theory [12,28,29]. In the cases of sliding surfaces, the contact period between points on the two surfaces is longer when the surfaces slide slowly than when they move rapidly. Consequently, if the slide of one surface over another slows down, friction increases. This situation favours the stick-slip phenomenon. However, laboratory tests have produced the unexpected findings that, at extremely slow speeds, the situation gets reversed: as friction increases, the sliding velocity also increases. This observation is related to the

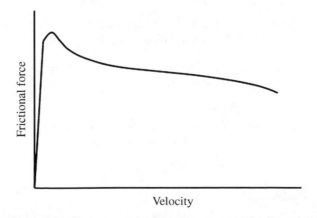

Figure 3. Effect of velocity on friction.

'creep' effect. All materials slowly change shape even under moderate force. An increase in force will increase the rate of creep. Thus, in the case of surfaces sliding very slowly over each other, an increase in frictional force may produce a perceptible acceleration of the slide in the form of creep of one surface past the other. The limit of speed attained by the creep mechanism varies with the material because soft materials creep faster than hard ones. The creep of steel, for example, is so slow that it cannot be observed. Lead, on the other hand, can be made to slide by creep at speeds up to a millionth of a centimetre per second.

2.2.1. Mechanism of friction

If a block of mass, m, with initial velocity, v_0, is projected along a horizontal table, it eventually comes to rest. It can be concluded from this experiment that while it is moving, it experiences an average acceleration a that points in the direction opposite to its motion. This is the frictional force, whose average value is ma, which the table exerts on the sliding block according to the Newton's second law of motion. Whenever the surface of one body slides over that of another, each body exerts a frictional force on the other, parallel to the surfaces. The direction of frictional force on each body is in a direction opposite to its motion relative to the other body. Even if there is no relative motion, frictional forces will exist between two surfaces contacting each other. The other type of friction is rolling friction, which occurs when a wheel, ball or cylinder rolls freely over a surface. Coefficients of sliding friction are generally 100 to 1000 times greater than that of rolling friction for corresponding materials. This phenomenon was realised historically with the transition from sledge to wheel [30].

The frictional forces acting between surfaces at rest with respect to each other are called forces of the static friction. The maximum force of static friction will be the same as the smallest force necessary to start a motion. Once motion is started, the frictional forces acting between bodies usually decrease so that a smaller force is necessary to maintain a uniform motion, which is called the force of kinetic friction. The ratio of the magnitude of the maximum force of the static friction to the magnitude of the normal force is called the coefficient of the static friction, and the average is given by

$$\mu_s = \frac{F_s}{N} \text{ or if the force is a function of time, } \frac{\int_0^t F_s dt}{\int_0^t N dt}. \tag{1}$$

The ratio of the magnitude of the force of the kinetic friction to the magnitude of the normal force is called the coefficient of the kinetic friction and the average is given by

$$\mu_k = \frac{F_k}{N} \text{ or if the force is a function of time, } \frac{\int_0^t F_k dt}{\int_0^t N dt}. \tag{2}$$

The coefficients of the static and kinetic friction are dependent on the magnitude of the normal force and the type of materials. Thus, the frictional force is linearly proportional to the normal force.

2.3. Mechanism of stick-slip

The breakdown of the third law of friction (the variation of frictional force with velocity) is responsible for stick-slip phenomenon. In order to understand this phenomenon better,

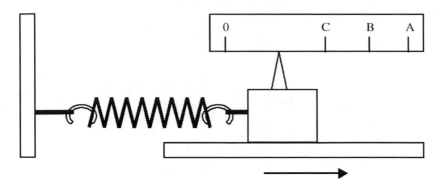

Figure 4. Experimental apparatus used for friction measurement.

let us give the following illustration. Suppose a block is attached to an anchored spring and placed on a longer slab, which was set in motion at a low speed as seen in Figure 4. At first, the block is dragged along on the moving slab: it will not be held back by the spring. It will slide on the slab, until the spring's pull is equal to the static frictional force. The pull of the stretched spring reaches that value when the block arrives at the point A. At this point, the block begins to slip on the moving surface. As soon as it does, the lower kinetic coefficient of the friction takes over, and the block slides rapidly towards the left. When it has moved back to the point C, it comes to rest. Here the higher static coefficient is effective in the motion and the block again sticks to the surface and is dragged to the point A. Then, it slips back to point C. This is a simple laboratory demonstration of the stick-slip phenomenon proposed by Bowden and Leben [12,30,31]. Morton and Hearle [23] stated that this apparatus was the best general method for the fundamental study of friction. Most of the apparatus developed to measure the friction of fibres use some modification of this methodology.

At point B on the scale, halfway between points A and C, the pull of the spring is equal to the kinetic coefficient of friction. If the static coefficient were the same as the kinetic, the block would be dragged to this point and then stay there, sliding on the moving slab beneath it. As it is, the block oscillates sticking and slipping by turns about this position. The situation is more complicated than it is described here. In fact, during the motion, the friction coefficient varies with changes in the sliding velocity. Whether stick-slip may or may not occur can be determined in any given situation simply from the direction in which this relation changes. If one wants to record the friction forces versus time in the experiment described above, with an appropriate signal-recording device, two different types of curves may be obtained depending on the speed of the slab. If the speed of the slab is low, the shape of the curves looks like a saw-tooth shape, although at the high speed it is more likely a sinusoidal, as given in Figure 5.

Most investigators [12,20,21,28,32–34] now agree that friction arises from the adhesion of molecules on the surfaces in contact with each other. The bond between the surfaces may be so strong at some points that tiny fragments are torn off one and stick to the other. In order to understand the rheological behaviour of stick-slip phenomenon, radioactive tracer materials have been used in a laboratory experiment [12]. If the end of a radioactive rod is rubbed along a flat surface, small particles are transferred and make the surface radioactive. This would be an excellent experiment for showing the stick-slip phenomenon. This experiment has been performed as follows: A piece of photographic film is laid on the surface rubbed with the rod that has been made artificially radioactive. After it has been

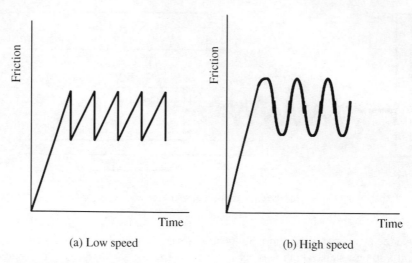

(a) Low speed   (b) High speed

Figure 5.   Stick-slip at different speeds.

exposed for several hours to the radioactive track left by the rod, the film is developed. The image of the track turned out to be not a continuous line but a series of spots. The sliding rod end stuck and slipped, leaving a considerable amount of material where it stuck and very little where it slipped.

2.4. *Friction in cylindrical surface*

The fundamental equation of friction can be written in a different form with the following analysis for cylindrical surfaces. Consider a yarn passing round a fixed guide (the guide should be fixed, otherwise the rolling friction and the kinetic energy of the guide should be taken into account), as shown in Figure 6. T_1 is the incoming tension and T_2 is the output tension, θ is the angle of contact and μ is the coefficient of friction. The output tension T_2 must be increased by an amount necessary to overcome the frictional resistance. For the sake of simplicity, the effect of twist blockage is neglected in the analysis. The change in the normal force due to tension in the yarn is given by

$$dN = T \sin\left(\frac{d\theta}{2}\right) + (T + dT) \sin\left(\frac{d\theta}{2}\right).$$ (3)

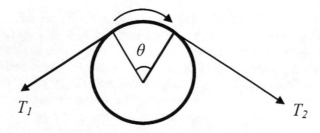

Figure 6.   Friction in yarn passing round rod.

Since $d\theta$ is small, the term $\sin(d\theta/2)$ can be approximated by $d\theta$. Neglecting the higher-order terms gives the following relation:

$$dN = T\,d\theta. \tag{4}$$

The increase in the tension is due to frictional drag. Therefore, the fundamental equation of friction can be applied as follows:

$$dT = \mu T\,d\theta. \tag{5}$$

Substituting Equation (4) into Equation (5) gives

$$dT = \mu T\,d\theta. \tag{6}$$

This equation is a separable differential equation. Therefore, using the separation of variables technique and appropriate limits of the integration, Equation (6) can be given in the following form:

$$\int_{T_1}^{T_2} \frac{dT}{T} = \int_0^\theta \mu\,d\theta. \tag{7}$$

The solution of Equation (9) gives the following relation:

$$T_2 = T_1 e^{\mu\theta}. \tag{8}$$

This is known as the Capstan Equation (also known as the Euler's Equation). The result of this friction force is only precise whenever the coefficient of friction is constant. Otherwise, the Capstan method will not give an accurate result.

2.5. Friction in textiles

Friction exists as in the form of fibre-to-fibre or fibre-to-machine parts, and is a common phenomenon in textile operations. Yarns are being constrained to move from process to process by guide elements [35,36]. Although the fibre friction is an important phenomenon in every stage of textile manufacturing from opening to weaving, it can be said that it is more critical in drafting operations (Figure 7). De Luca [37] reported that in roving, drafting force and fibre frictions were more important than torsion and bending rigidities. Therefore, it is appropriate to give a brief overview about friction in the drafting field before attempting to give a survey about friction models and measurements in fibrous materials.

2.5.1. Friction in drafting field

Mutual shifting of fibres relative to each other along the flow direction, which is the essence of roller drafting, is achieved when each fibre enters the drafting system with lower velocity, v_1, and leaves it with higher velocity, v_2. In the drafting field, some of the fibres have a velocity of between v_1 and v_2. For the sake of simplicity, it is assumed that fibres are fed into the drafting zone one by one. A fibre enters the system as an element of the first group moving with velocity, v_1. When the first fibre reaches the second pair of rollers, its velocity

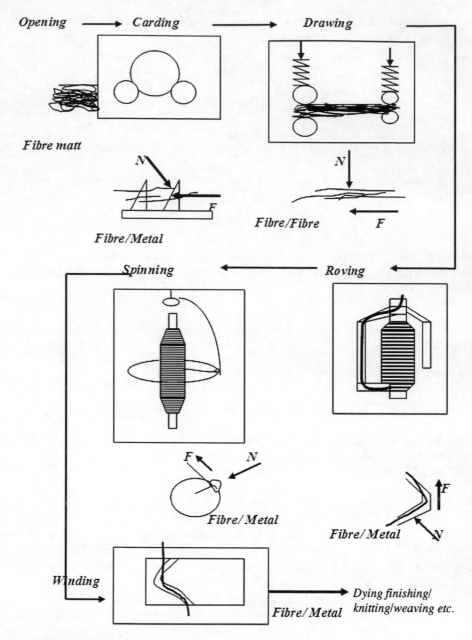

Figure 7. Friction in textile applications.

would be v_2, and consequently, will slide over fibres still in contact with the back rollers, moving at v_1. This would develop a friction force between two fibre groups.

When a force is exerted over on assembly of rigid bodies, the force can be spread within a certain space and experienced by numerous bodies that contact each other. In Figure 8, for instance, the roller pressure, P, exerted at a single point, is spread over the entire length of the fibre that transmits the force to other fibres. Therefore, in a sliver composed of rigid fibres, the pressure exerted by the rollers should extend far away from them. In the case

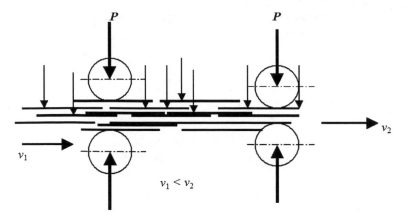

Figure 8. Friction zone.

of perfectly flexible fibres according to their geometry, the area in which the pressure is exerted on the fibres is confined to the region directly in contact with, and deformed by, the rollers. In reality, fibres are neither rigid nor are they perfectly flexible. Therefore, lateral pressures over the fibres are exerted at some short distance away from the deformation zone between the rollers. The important factors that determine the characteristics of the friction field are given as follows [13,38–40]:

2.5.2. Factors affecting the friction field

- *The pressure of the rollers*: Generally, the higher the load, the higher would be the intensity at every point of the friction field. Since the minimum intensity is reached at a greater distance from the middle point, the friction field is both longer and wider.
- *Fibre arrangement*: The cross section of sliver does not have a definite form. A 'quasi-circular' section can be defined for a sliver cross section in which the number of fibres decreases from the middle region to both sides and so does the pressure. It is also possible to define a 'quasi-rectangular' section in which the fibres are uniformly spread across their width, and the pressure is uniform. The uniform pressure case can be obtained by using a condenser with a special profile. Non-uniformity of the fibres' arrangement across the sliver can entail variation of the intensity of the friction field, especially when the number of fibres is low or they are spread over a large roller [41,42].
- *The radii of the rollers*: The pressure over the fibres is exerted by the geometry of the rollers. In the region of minimum distance between the roller surfaces, the strand is subject to a maximum deformation and the pressure exerted has a maximum value. It decreases both ahead and behind as the roller surfaces depart from each other. Therefore, as the diameter of the roller increases, the friction field also increases.
- *Density of sliver*: The effect of this factor is not too strong in comparison with the others. However, since an increase in the number of fibres entails a greater thickness, contact of sliver with the rollers extends over a larger surface. As a result, length of the friction field increases, although its middle intensity decreases. It is important to note that friction field variation is much smaller than variation in the corresponding number of fibres.
- *Characteristics of roller surface*: A rigid fabric does not comply with variation of the fibre density across the sliver. As such, most of the roller pressure is concentrated

in those regions where the fibre density is higher, although the region with lower density is deprived of pressure. To level the pressure over fibres across the sliver, an elastic cover is appropriate. Similar effects may also be obtained along the sliver, that is, the elastic cover extends the direct contact between the lower and upper rollers, and levels the pressure along the friction field.

3. Friction force in friction field

In the drafting zone, a fibre is subjected to a friction force due to contact with other fibres and rollers. For simplicity, we presume a fibre to be straight, inextensible and move along its length with the surface velocity of the rollers. Furthermore, it is also assumed that the contact of a fibre with surrounding fibres and rollers is continuous and their pressure is normal to the fibre. Then the total friction force over the fibre developed in a particular friction field is given by

$$F = \int_{x_t}^{x_l} dF(x) = \int_{x_t}^{x_l} N_1(x)dx, \tag{9}$$

where x_t is the trailing end and x_l is the leading end. There are six different possible arrangements, as shown in Figure 9, for a fibre in the friction field. When fibres are in the friction field, the integration limits are from the tailing end to the leading end. If fibres are entirely out of the friction field, then no possible friction force can be developed in the friction field. In this case, the fibres keep moving due to the adhesion forces with the adjacent fibres. The magnitude of the adhesion force is mostly related to the static coefficient of the friction [39,40].

In Foster's 'two velocities hypothesis', a fibre enters the friction field with the velocity of the back roller until it reaches the front roller; then, its velocity will be the velocity of the front roller. The number of fibres at the back roller is n_1 and that of the front roller is n_2. The static and kinetic frictional forces developed in the friction field would be the following, with indices one representing the back roller and indices two representing the

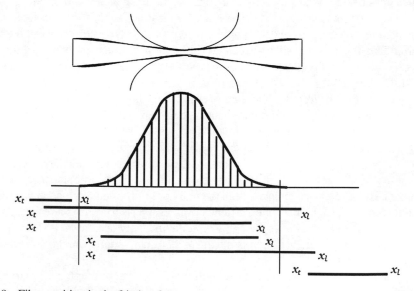

Figure 9. Fibre position in the friction field.

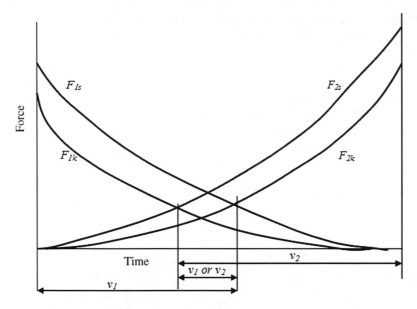

Figure 10.   Withdrawal and friction forces acting on a fibre.

front roller. Because μ_s is greater than μ_k, the following inequalities are always satisfied:

$$F_{1s} = \mu_s \int_{x_t}^{x_l} n_1(x)N(x)dx, \tag{10}$$

$$F_{2s} = \mu_s \int_{x_t}^{x_l} n_2(x)N(x)dx, \tag{11}$$

$$F_{1k} = \mu_k \int_{x_t}^{x_l} n_1(x)N(x)dx, \tag{12}$$

$$F_{2k} = \mu_k \int_{x_t}^{x_l} n_2(x)N(x)dx, \tag{13}$$

$$F_{2s} > F_{2k} \; and \; F_{1s} > F_{1k}. \tag{14}$$

At the beginning, the majority of fibres are moving with v_1, thus both F_{1s} and F_{1k} are higher than F_{2s} and F_{2k}, respectively. Then, the first two forces decrease and eventually vanish, although the last two become gradually higher and higher, as shown in Figure 10. Consequently, when a fibre reaches the higher velocity, it keeps moving and does not return to the slower velocity [43–46].

3.1. Friction of textile fibres

The values of static and kinetic friction coefficients have an important meaning in textile applications. In general, it is desirable to have a low fibre friction during drawing process in order to avoid a high resistance to the drafting forces. However, a certain level of friction is also required for a better cohesion in the spinning processes. As mentioned earlier, this is the force giving strength to the yarn and is also primarily responsible for adherence of fibres during processing. Therefore, it is a requirement to have a friction force among the fibres themselves as well as with the surface of the machine parts for an optimum process. As textile science introduces new synthetic fibres, the need to establish a good measurement method for the fibre friction is rising rapidly.

Although it is possible to analyse the frictional properties of each fibre type separately, it would be more convenient to discuss them as two groups, namely, natural and synthetic fibres. It would be easier to understand the properties of a homogeneous material than those exhibiting local variations or anisotropy. Synthetic fibres are, in contrast with natural fibres, generally homogeneous to a first approximation in their surface structure. The frictional characteristics of fibres are as follows.

3.1.1. *Friction of natural fibres*

As is well known, the properties of natural fibres are highly variable, so are their frictional characteristics. Natural fibres exhibit local variations along their axis and the properties of the fibre depend on the testing direction. Frictional characteristics of natural fibres (wool, cotton, long vegetable fibres etc.) have been studied by many authors [47–61]. It was found that frictional properties are related to their surface characteristics and, further, their histological structure. Furthermore, the histological structure of natural fibres is an important factor for the evaluation of their frictional characteristics.

3.1.1.1. *The friction of wool and fur fibres.*
The direction of rubbing of fibres or yarns will influence the coefficient of the friction. This can be easily observed with wool fibres. It was reported that the friction of wool fibres was dependent on the direction towards which it was pulled. This is known as the directional frictional effect (DFE), and the coefficient of friction is different depending on the pull direction [25,62–64]. If one examines a wool fibre, it will be noticed that the cuticle cells form an overlapped structure. The exposed part of each scale points towards the tip end of the fibre. It is due to this cuticular morphology that inter-fibre friction of wool depends on the direction of rubbing. This directional effect is the basic cause of the felting and has an important practical value for wool fibres [52,65–68].

3.1.1.2. *Cotton fibres.*
The morphology of cotton fibres can be described as a ribbon with distinct twists, commonly known as spirals or convolutions, about its longitudinal axis with the helix [69]. Contacts among cotton fibres occur at the tips of the convolutions. This feature has an effect on the frictional characteristics of cotton fibres. Hood [70] reported that the mercerisation of cotton fibres caused a significant increase in the inter-fibre friction. His findings were confirmed by Mody and his co-worker [71,72]. Although no explanation was provided for this mechanism, it is well known that the mercerisation makes the cross section of the fibre circular due to swelling. This gives more contact points along the fibre axis. Terms, such as cohesion, clinging power and stick-slip need to be considered when discussing the frictional characteristics of cotton fibres. In most of the analysis of friction, these phenomena have not been explicitly included in the model. They are included implicitly by considering the coefficient of friction, the friction index or any other type of intrinsic coefficients [21,73–75]:

- *Cohesion*: This force holds the fibres together during the manufacturing of sliver, roving and yarn. This is more important when fibres are in the form of a sliver with almost no twist on it. The cohesion is the only force holding the fibres together with clinging power in the sliver. Cohesion is the result of the inter-fibre friction, twist or the combination of both of them. The alignment and the thickness of fibres have a great influence on the cohesion among the fibres. Generally, finer and combed fibres give a higher degree of cohesion than coarse and regularly carded fibres, largely due to the alignment of the fibres and the true area of contacts. In cotton fibres, the interlocking of hair convolution is another reason for cohesion. As the cohesion increases, the evenness and the strength of the yarn also increases.

- *Clinging power*: This is the force to hold adjacent fibres together. The term 'clinging power' was first introduced by Adderley [47] to describe the characteristic behaviour of cotton fibres. The higher the normal pressure, the more the number of contact points, thus the higher would be the clinging power. The clinging power is also affected by the fibre fineness, convolution, regularity, cleaning ratio and the fibre-blending ratio. Hood [70] showed that the resisting frictional forces were related to the presence of convolutions [76].
- *Scroop*: When a fibrous material is compressed by hand, it is possible to characterise the material from its sounds. It is usually associated with silk, but it can also be observed in certain cellulosic fibres, yarns and fabrics by applying suitable finishes to them. More interestingly, scroop is also affected by the coefficient of frictions. It is directly related to the high coefficient of static friction relative to the dynamic friction.
- *Stick-slip*: This very complex phenomenon can be explained simply by the name itself. Fibres stick to each other because of friction and when a little force is applied such that the force exceeds the static friction, the fibre slips until it finds another sticking point to which it can hold.

3.1.1.3. Other natural fibres. The frictional characteristics of jute, flax, ramie and other fibres are governed by their special surface properties. Study of friction for these fibres showed that they have a practical importance for their spinnability. Andrews [77] and Louis [78] have reported that the spinnability of milkweed fibres is limited by their stiffness and soft surface. Lombard and his co-workers [79] conducted an experiment to increase the frictional coefficient and reduce the stiffness of milkweed fibres. From the work of Andrews, Louis, Lombard et al., it is necessary to have an appropriate coefficient of friction; otherwise, fibres cannot be spun efficiently.

3.1.2. Friction of synthetic fibres

It has been shown that the surface geometry of natural fibres is complex. Synthetic fibres, on the other hand, have a relatively simple and uniform surface geometry compared with natural fibres. Due to the relatively homogeneous structure of synthetic fibres, most of the properties of synthetic fibres are quite predictable, including their frictional characteristics. The frictional behaviour of synthetic fibres could be represented by the fundamental equation ($F = aN^n$) proposed by Howell [33,80] and confirmed by some other investigators [33,67,81,82]. Robins, Rennell, and Arnell [83] stated that the frictional characteristics of cleaned polyester fibres were consistent with the adhesion theory of friction proposed by Bowden and Tabor [28,29].

4. Theories on friction of fibrous materials

During the manufacturing of yarns and subsequent processing of synthetic fibres, yarn tension should be controlled properly. There are two types of forces involved in the control of tension: one is the driving force and the other is the frictional force. In order to establish better control of the process, the nature of the friction should be understood properly. Fibre friction is of importance simply because it is the primary property responsible for the mutual adherence of fibres during processing. Although the fundamental theory of the friction was given for plastic materials, no such concepts of friction exist for fibrous materials and polymers. The relationships, which describe the frictional characteristics of those materials, are purely empirical.

4.1. Adhesion Shearing Theory

As has been discussed earlier, most of the surface and morphological properties of the materials were included in the friction equations as a constant term called the coefficient of friction. The meaning of the parameter, μ, was given in the well-known 'Adhesion Shearing Theory' [28,29] for metal surfaces. According to this theory, contacts occur only at the tips of the asperities (Figure 11). When two bodies are brought into contact, the area of each asperity in the contact region is initially small and the pressure is high. Consequently, a plastic flow occurs in the asperity that leads to an increase in the area of contact and a decrease in the pressure. This continues until a condition at which no further plastic flow occurs and the asperity begins to deform elastically. This condition is the equilibrium condition and has been shown in Figure 12 for metal and fibrous materials. If the pressure is higher than the yield pressure, plastic flow occurs and gives an increase in the area and thus a decrease in the pressure. The asperities of the object deform until the pressure decreases to the value where the pressure could be supported elastically. Bowden and Tabor [28,29] have stated that adhesion develops at the points of contact between two bodies. In order for sliding to occur, the adhesion developed between bodies must be sheared. If A is the total area of sheared and S is the average shear strength of the weaker material, the adhesion component of the friction given by

$$F_A = AS. \tag{15}$$

The discussion of the friction mechanism for metals would appear to be quite valid as far as the behaviour of a solid (in a sliding and rolling process) depends on their deformation and the surface properties. Although the formula of a general theory does not promise to be rewarding for the viscoelastic and fibrous materials, the common concepts of the adhesion theory of metallic friction have been extended to include polymer friction. Schick [84] reported that during sliding of one polymer surface against another, strong adhesion occurred and fragments sheared off from one polymer were observed adhering to the other polymers. Adams [85,86] measured the friction and apparent contact area of nylon on glass and concluded that the tangential force had a little effect on the contact

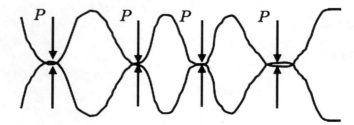

Figure 11.   Distribution of load over contact between two objects.

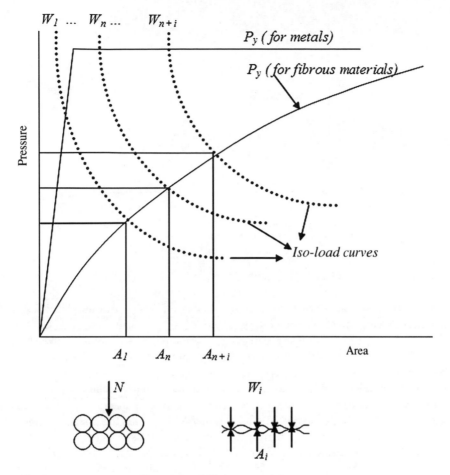

Figure 12. Pressure-area curve for metal and fibrous materials.

area. He also stated that the shear strength of the contact area between nylon and glass increases with pressure. Thus, it seems that there is a reasonable agreement between the adhesion theory of metal and polymer frictions. However, it should be noted that there is a difference between the natures of deformation of these two materials, which is as follows: metals generally exhibit plastic deformation, yet polymers exhibit elastic deformation over certain ranges of loads. Because of this viscoelastic deformation, there is a hysteresis loss with polymeric substances [87]. This viscoelastic deformation influences the total frictional force; therefore, the classical law of friction no longer holds good. With many solids, the deformation of irregularities is elastic rather than plastic or it may follow an intermediate law. It was reported that the area of contact grew nearly proportionally with the load [11]. This was indeed confirmed by Bowden and Tabor [28,29] who reported that the frictional force was approximately equal to the product of the area of contact and the shear strength of the plastic bulk. Many polymers show viscoelastic behaviour and are easily deformed; hysteresis fails to contribute a large fraction of the total sliding resistance in the case of metals [29,88]. Therefore, fibres and polymers do not show a linear relationship with the normal force, as shown in Figure 13.

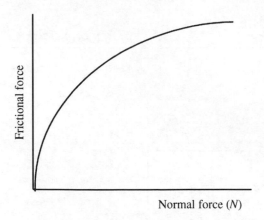

Figure 13.   Frictional forces vs. normal forces in fibres.

Adhesion forces between two surfaces are regarded as coming from two main sources, that is, electrostatic and van der Waals forces. If the materials in contact have different polarity, electrostatic forces are established between two surfaces. Besides the adhesion, the deformation mechanism also has an important effect when two surfaces slide against each other and plough out a series of corrugations. The principle of deformation, given by many authors [28,29,44,89–91], is as follows: input energy into the polymer ahead of the asperity and lose energy at the rear of the asperity. It was reported that the net loss of energy is related to the input energy and the loss properties of the polymer at a particular temperature. More interestingly, theories about the deformation mechanism have stated that the deformation term involves the bulk properties of the polymer rather than the special condition of surfaces [20]. Since the friction of rigid polymeric materials is generally composed of the adhesion and the deformation mechanisms, many researchers have investigated the importance of these mechanisms [20,22,23,92]. The adhesion mechanism is more related to the surface chemistry, namely, attraction between two surfaces. As a general summary of the experimental investigations, the following features represent the frictional properties of rigid polymers:

- The friction is affected not only by the magnitude of the load but also by the geometry of the surfaces. Generally, smooth surfaces are preferred for less frictional forces.
- The friction of rigid polymers depends on the sliding speed and the temperature in a manner that reflects their viscoelastic properties [20,93–95]. Depending on temperature, the contribution of the force coming from the adhesion and deformation mechanisms can be changed. Theoretically, the viscoelastic properties of materials have not been included into most of the frictional models. However, the material-dependent constant term inherits the viscoelastic properties of materials.
- The friction–velocity–temperature curves of polymers in a glassy state can be transformed by a suitable series of shifts into a single master curve by 'Arrhenius factor'. This factor is dependent on the viscoelastic properties of the polymer. Therefore, it can be determined from viscoelastic measurement of the polymer [20,96].
- In the friction process, the net energy loss may be in the form of elastic hysteresis. Most of the lost energy is converted into heat that raises the temperature of the surface. The rise in temperature also affects the viscoelastic properties of the material, and it will change the friction–velocity–temperature relationship.

- The frictional force for a rigid polymer can be given as a summation of the adhesion and deformation components. The adhesion component is a function of the contact area, and the deformation is related to the normal force.
- The relative importance of these two components has been found to be influenced by the nature of the contact surface, temperature, sliding speed, thickness of the polymer and properties of the lubricant used in the process [20,28,96].

4.2. Friction studies in textile fibres

The general characteristics of textile fibres are their fineness, flexibility and interesting morphological structures. All of these properties have an effect on the frictional characteristics of fibres. Gralen [94] conducted an inter-fibre friction test and showed that the classical law of friction was not valid for fibrous structures. The frictional force generally decreases with increasing load. Furthermore, the coefficient of friction was not a constant, and there was not a simple expression for it. Therefore, the evaluation of friction for textile materials is not an easy task. Mercer and Makinson [53], Makinson [51,97] and Howell [92] confirmed the findings of Gralen [94] and proposed a simpler relationship ($F = aN^m$), which was claimed to better fit the experimental data of fibrous materials.

4.2.1. Pressure area relation in structural model of friction

As mentioned earlier, the contact area in a viscoelastic system is pressure dependent; therefore, the magnitude of pressure is an essential factor for true area in contact. Pressure P related to area is given by the following equation:

$$P = KA^a. \tag{16}$$

Constant K, the stiffness or the hardness factor of the material, and parameter a related to the shape of the objects, define the nature of the relationship [98]. Depending on the values of K and a, the relationship between the pressure and the area is given in Figure 14.

Gupta and El Mogahzy [98,99] grouped the factors affecting the value of friction in fibres to fall in two main categories. One governs the morphology of the contact, and the other the mechanical properties of junctions. According to the Bowden and Tabor's Adhesion Shearing Theory [28,29], as explained above, junctions are formed at the points of real contact, which must be sheared before sliding. Therefore, the frictional force F is given by the product of the true area of contact A and the bulk of specific shear strength of the junction S, as stated in Equation (15). For materials deforming plastically, such as ductile metals, the area A is given by the ratio of the normal force N to the yield pressure P_y of the material. Therefore, Equation (15) can be re-written as follows:

$$F = AS = \left(\frac{N}{P_y}\right)N = \mu N. \tag{17}$$

The coefficient of friction is a material property in this classical equation. Based on the analysis performed above, Gupta and El Mogahzy [98,99] and El Mogahzy [20] developed a structural model for acrylic and polypropylene yarn, as follows:

$$F = SC_M \left(\frac{1}{K}\right)^n m^{1-n} N^n. \tag{18}$$

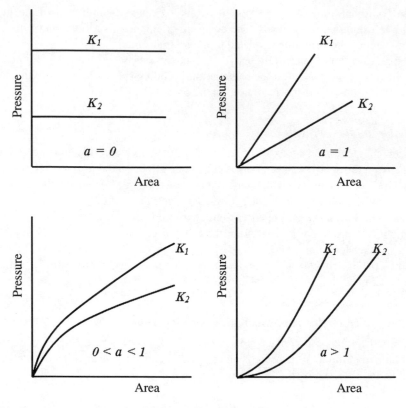

Figure 14. Pressure area curves for different values of a and K, $K_1 > K_2$.

The terms in the equation were divided into two categories. First, the morphology of contact is given by the number of contacts between bodies (m) and the nature of the stress distribution on the contact region. This can be described by the model constant (C_M), whose value is 1, for the uniform stress distribution and between 0.92 and 1 for the other distributions. Second is related to the mechanical behaviour of the junctions and given by the values of constants K and a. However, the latter is related to the pressure area relationship, given in $P = KA^a$, and the value of specific shear strength of the junction, S. They stated that the parameter K represents the stiffness of the material and a is the shape of the pressure curve. El Mogahzy [20,98–100,] also showed that the empirical constant n could be related to the theoretical constant a by the relation

$$a = n^{-1} - 1. \tag{19}$$

The main weakness of the above model is that it depends on the number of asperities. It is not an easy task to define the correct number of asperities for any materials. The second question arises with the distribution of the stress over the asperities. In their model, it was said that C_M is the model constant and depends on the stress distribution. El Mogahzy [20] assumed the stress distribution related to the cross section of the fibre, i.e. if the cross section of the fibre was circular, the stress distribution was accepted as uniform. He has also conducted his research by using triangular and trilobal cross sections corresponding to conical and spherical stress distribution, respectively. In the model assumptions, El

Mogahzy reported that contacts only occur at the tips of the asperities existing on the surface of the body. Therefore, it is illusory to expect that the stress distribution would be the only function of the fibre cross section. The model constant is more likely a shape factor of the fibres. The distribution parameters were defined in their model for uniform, spherical and conical stress distributions. Regardless of the stress distribution, the model parameters m, n, K and normal force N were found identical. In the model proposed, the coefficient of friction is given by

$$\mu = SC_M K^{-n} m^{1-n} N^{n-1}. \tag{20}$$

As seen, the new equation did not change the fundamental meaning of the coefficient of friction, as a ratio of the friction force to the normal pressure $\mu = F/N$. However, instead of being a constant factor, as in Amontons' equation, it has become a variable dependent upon N. Comparing the previous two equations with Equation (36), $F = aN^n$ yields the empirical constant a as follows (note that n is already in the model):

$$a = SC_M K^{-n} m^{1-n}. \tag{21}$$

With this model, it is possible to express the theoretical constant a in terms of a number of fundamental factors, some of which are related to the material properties and the rest related to the morphology of the contact points and stress distribution.

El Mogahzy and Broughton [21,101] later expanded the model given by Gupta and El Mogahzy [98,99] and explained the concept of the friction profile as well as measuring technique. In their paper, an extensive theoretical analysis of the frictional coefficient of cotton fibres was given as follows: They considered two different cases of fibres contacting the same reference surface shown in Figure 15. In case A, the number of fibres in contact is n_1, and in case B, the number of fibres in contact is n_2. They assumed that the total normal force applied to the fibre assemblies in both cases were constant and had the magnitude of

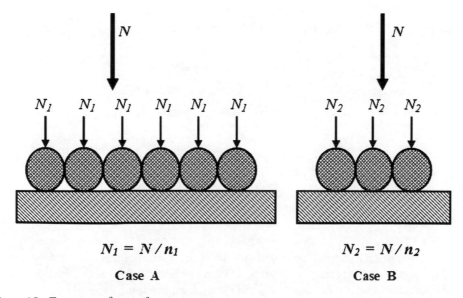

$$N_1 = N/n_1 \qquad\qquad N_2 = N/n_2$$

Case A   **Case B**

Figure 15. Two cases of area of contact.

N. In order to simplify the analysis, all fibres were assumed to have identical dimensions. If the classical law of friction holds, the normal force per fibre will be $N_1 = N/n_1$ and constant for each fibre. Therefore, the frictional force per fibre is given by

$$f_1 = \mu N_1 = \mu \frac{N}{n_1}. \tag{22}$$

The total frictional force is the summation of the forces over the fibre assembly, which is given by

$$F_A = \sum f_1 = \mu \frac{N}{n_1} n_1 = \mu N. \tag{23}$$

Similarly, the total frictional force for case B is also $F_A = \mu N$. Thus, $F_A = F_B$. This represents the basis for Amontons' law of friction, that is, for a given normal force; the frictional force is independent of the area of contact. The relationship $F = aN^n$ was also considered for each of the two cases. In case A, the frictional force per fibre is given by

$$f_1 = a_1 N_1^n = a_1 \left(\frac{N}{n_1} \right)^n. \tag{24}$$

The total frictional force for case A is then given by

$$F_A = \sum f_1 = n_1 a_1 \left(\frac{N}{n_1} \right)^n = a_1 n_1^{1-n} N^n. \tag{25}$$

Similarly, the total frictional force for case B is given by

$$F_B = \sum f_2 = n_2 a_2 \left(\frac{N}{n_2} \right)^n = a_2 n_2^{1-n} N^n. \tag{26}$$

Furthermore, the ratio of F_A and F_B was taken and the result was given as follows:

$$\frac{F_B}{F_A} = \frac{a_2}{a_1} \left(\frac{n_2}{n_1} \right)^{1-n} \quad or \quad F = Ca(n_f)^{1-n}. \tag{27}$$

In this model, C was defined as a model constant. Equation (27) was related to the model given by Gupta [98,99]; therefore, it can be given in the following form:

$$F = a_0 (n_f)^{2(1-n)}. \tag{28}$$

The term a_0 was considered as a friction index and defined by

$$a_0 = CSC_m K^{-n} \delta^{1-n}. \tag{29}$$

It was reported that the friction index was strictly related to the deformation of the junction during friction. A high value of a_0 is an indication of high shear strength of the junctions or a low hardness of the junctions under lateral stresses. From the experimental data, the empirical values of a_0 and n can be estimated [21]. Yet, the model depends upon the number of contact points. They assumed that the number of contact points was a linear function of the number of fibres by introducing

$$m = \delta n_f. \tag{30}$$

Since fibre length does not have a uniform distribution, the number of contact points may not exhibit linearity. Thus, the number of contact points may give inaccurate readings for the fibre friction test.

Furthermore, the work of El Mogahzy and Broughton [21] is a continuation of the work of Gupta and El Mogahzy [98,99] in a sense of theoretical analysis. As seen on the analysis, the equations provided for characterising the frictional phenomenon do depend on the characteristics of junction, which cannot be obtained with elementary analysis. However, the idea of introducing the friction indices is indeed a useful tool for the analysis of fibre frictions. The evaluation of the fibre friction in this way requires a great deal of experimentation because the friction index should be determined for each type of fibre separately. In the case of natural fibres, such as cotton and wool, the scenario is even worse because the friction index tends to be changing from year to year.

A characteristic frictional behaviour often exhibited is stick-slip motion that occurs when the friction force drops suddenly at the initiation of sheared motion. Stick-slip motion may have varied characteristics depending on applied load, sliding velocity, surface roughness and humidity. Classical investigation of stick-slip motion was performed by many researchers including Rabinowics [3,5,7,9,12,102–104]. Rough surfaces may apparently generate a stick-slip motion, as the asperities of one surface slide over those of the opposite surface. Surface roughness-induced stick-slip motion shows an irregularity of stick-slip amplitude, which depends on heights, slopes and elastic modulus of asperities. This explanation, given by Gao [105], indicates that stick-slip motion can be caused by only surface roughness rather than the intrinsic interactions of lubrication media. These phenomena cause a large static friction force at the beginning of sticking. Gao and Wilsdorf [105] reported that an asperity surrounded by a liquid film causes the formation of a meniscus due to surface energy that generates an attractive force. The sticking phenomenon may be affected by the attractive force generated by the liquid used as a lubricant.

Yoshizawa, McGuiggan and Israelachvili [106] did not agree with Gao [105] on the effects of a lubricant to the stick-slip motion. They were concerned with the attractive forces due to molecular bonding between surfaces to explain stick-slip motion. They stated that adhesion and stick-slip motion come from the shear thinning effect of the lubricant. The force to trigger the motion is higher than the force to maintain the motion after the initiation of motion. Reduced friction force with sliding velocity causes surface motions of a periodic nature. Therefore, the motion is caused by the intrinsic behaviour of the lubricant between two surfaces, not direct interactions of surface asperities. Jang and Tichy [107] have also agreed with Yoshizawa, McGuiggan and Israelachvili [106] about the stick-slip phenomenon affected by the attractive force generated by the lubrication. In the work of Gupta and El Mogahzy [98,99] and El Mogahzy [20] and Broughton [21], they have founded their model on the direct interactions of surface asperities. It seems there is a contradiction between the model proposed by the latter authors and Yoshizawa, McGuiggan and Israelachvili [106]. However, it should be kept in mind that Gupta and El Mogahzy did not consider the effects of a lubricant in the fibre assembly. In the case of using a lubricant on the surfaces, the asperities are covered by that lubricant; therefore, the interaction will occur among lubricant molecules.

4.2.2. Capstan method in structural model of friction

Since most of the stationary surfaces present a radius of curvature to the yarn, the Capstan equation has been a subject of many theoretical and experimental studies in the friction of fibrous materials [108–111]. The Capstan equation is limited due to the fact that it does not include any critical parameters of the yarns or fibres, such as speed, diameter and bending of the yarns or fibres. Martin and Mittelman [112] have stated that their experiment disagreed with Amontons's law. Later, Breazeale [88] confirmed the deviation from the Amontons' law. Gonsalves [113] tried to measure the coefficient of rayon fibres and conducted his test by using the Capstan method. He reported that the coefficient of friction was not a constant, but varies with T_1. However, he also claimed that μ could be treated as a constant for a small value of the wrapping angle. Therefore, the Capstan equation is not an adequate solution for the friction of fibrous materials. For example, when incoming tension T_1 is zero, T_2 is also zero, regardless of the value of the wrap angle.

Olofsson [114] further pursued the theoretical study for the ratio of (T_2/T_1) dependence on T_1. He suggested that the results of Martin and his co-worker [112] could be a reasonable fit if Amontons' law was modified according to the following equation:

$$F = \mu_0 N + \alpha A,$$
(31)

where μ_0 and α are constants and A is the contact area. Then, taking the routine derivation, he proposed the following equation in which β is given as the model constant:

$$T_2 = T_1 e^{\mu_0\theta} + \beta \frac{e^{\mu_0\theta} - 1}{\mu_0}.$$
(32)

Howell and his co-worker [33], Howell [80,92,115] and Guthrie [81] modified the Capstan equation for a fibre or yarn wrapping a guide. They proposed the second most widely accepted and used equation, especially for viscoelastic and fibrous materials, as follows:

$$F = aN^n,$$
(33)

where a and n vary from material to material. Howell and Mazur [33] stated that the value of a and n were material dependent. Lincoln [34,82] agreed with the results reported by Howell and Mazur [33]. This alternative relation, giving the dependence of the frictional force on the normal force, has been the subject of many investigations [21,22,80,93,94,98,116–121]. Table 2 shows the experimentally obtained values for α and n for different fibres. The equation given above can be accepted to be the general law of friction for the fibrous materials. Howell [115] also studied the case where the coefficient of friction was not a constant (it was a function of the radius and the incoming tension). He accepted the following relationship between the coefficient of friction and the radius of the cylinder:

$$\mu = a \left(\frac{\rho}{T_0}\right)^{1-n}.$$
(34)

Then, the Capstan form of equation was given by

$$T_2 = T_1 e^{a\left(\frac{\rho}{T_0}\right)^{1-n}\theta}.$$
(35)

Table 2.   Value of α and n obtained experimentally.

Fibres–Fibres	α	n
Drawn nylon–drawn nylon	0.92	0.80
Undrawn nylon–undrawn Nylon	0.85	0.90
Viscose rayon–viscose rayon	0.49	0.91
Acetate–acetate	0.60	0.94
Acetate–nylon	—	0.81
Acetate–viscose rayon	—	0.90
Acetate–wool	—	0.92
Acetate–polyester	—	0.88

As seen from the equation, this model is also limited. It does not contain any parameters related to the material-bending properties; therefore, T_2 is zero whenever T_1 is zero. In the model, μ is a function of the radius and ρ of the friction surface. The experimental results from the works of Lyne [122], Olsen [123], and Baird and Mieszik [32] have confirmed that the coefficient of friction increases with the increase of the radius. The latter three experiments were performed by different sizes of guides from 1.7 to 25.4 mm with nylon and cellulose acetate yarns. However, some of the researchers [124,125] have disagreed with the conclusion of Howell [115] and the findings of Lyne [122], Olsen [123] and Baird and Mieszik [32]. Galuszynski [124] reported that the frictional force decreased with the increase of the guide diameter. He used the diameters in the range of 0.3 to 0.5 mm with wool, cotton and polyester yarns. McMahon [125], on the other hand, confirmed the results of Galuszynski and his co-worker [124] for the diameters in the range of 0.3 to 5.0 mm. As noticed, neither of the researchers reports the effect of the yarn speed. The disagreement with the proposed model arises mainly due to the use of different speeds, surface finishes and, most importantly, the bending properties of the yarn. As the diameter of the cylinder decreases, the bending moment of the yarn increases. McMahon [125] explained his results by using Kawabata and Morooka's [126] assumptions. Kawabata and his co-worker [126] used Bowden and Tabor's [28] assumption that the total friction force F is the sum of two components given by

$$F = F_A + F_D, \tag{36}$$

where F_A is the only component related to the surface contact between the guide and the yarn and F_D is related to other effects, such as yarn deformation. The contribution of the deformation component F_D is quite significant at small diameters. The developed theories about the frictional properties of viscoelastic materials were fitted for a wide range of applications. In order to accomplish this task, the Adhesion Shearing theory, developed by Bowden and Tabor [28,29], was widely used.

Howell [115] and Lincoln [82] have given another form of friction equation for the yarn passing around a peg. The radius of curvature and the angle of wrap were also included in this equation. This is an extension of the fundamental equation for fibrous materials $F = aN^n$ and given by

$$T_2 = T_1 \left[1 + (1 - n)a \left(\frac{\rho}{T_1} \right)^{1-n} \alpha \right]^{\frac{1}{1-n}} . \tag{37}$$

From this definition, the coefficient of friction can be derived as follows:

$$\mu = \frac{1}{\alpha(1-n)} \ln\left[1 + (1-n)a\left(\frac{\rho}{T_1}\right)^{1-n}\alpha\right]. \tag{38}$$

As seen in the above equation, the value of μ not only depends on radius and wrap angle but also on the magnitude of the incoming tension. Kowalski [127] developed an equation for the yarn friction with a solid body. He considered rheological parameters, flexural rigidity, dynamic viscosity, geometrical parameters and speed of the yarn in his model. He has given his model as follows:

$$T_2 = T_1 + c\dot\varepsilon t + \eta\dot\varepsilon\left(1 - e^{-\frac{lc}{\eta}}\right), \tag{39}$$

where $\dot\varepsilon$ is the yarn elongation given by

$$\dot\varepsilon = \frac{dT}{cdt}, \tag{40}$$

η is the relative viscosity of the yarn, c is relative flexural rigidity and l is the length of the contact line. The final derivation of the equation was given by

$$T_2 = H + \frac{nvT_1}{c\rho\alpha}\left\{\left[1 + (1-n)a\left(\frac{\rho}{T_1}\right)^{\frac{1}{1-n}}\alpha\right]^{\frac{1}{1-n}} - 1\right\}\left(1 - e^{-\frac{\rho dc_1}{v\eta}}\right), \tag{41}$$

where H is same as in Howell [115] and Lincoln's [82] equation (Equation (41)). The equation proposed by Kowalski [127] can be hardly considered practical due to its complexity. The model further depends upon numerous coefficients, some of which must be determined empirically. Additionally, the coefficients do not remain constant throughout the entire range of the surface.

Wei and his co-worker Chen [128] have proposed an improved Capstan equation for non-flexible fibres and yarns. They considered a conventional Capstan method, yet they added the shear force and bending moment acting on the yarn, as seen in Figure 16. If the inertia moment of the element is neglected, the equilibrium equations from the free body diagram of the yarn were given as follows:

$$\sum F_T = (T + dT)\cos\left(\frac{d\varphi}{2}\right) - T\cos\left(\frac{d\varphi}{2}\right) + (Q + dQ)\sin\left(\frac{d\varphi}{2}\right)$$
$$+ Q\sin\left(\frac{d\varphi}{2}\right) - dF\mu, \tag{42}$$

$$\sum F_n = (Q + dQ)\cos\left(\frac{d\varphi}{2}\right) - Q\cos\left(\frac{d\varphi}{2}\right) - (T + dT)\sin\left(\frac{d\varphi}{2}\right)$$
$$- T\sin\left(\frac{d\varphi}{2}\right) - dN, \tag{43}$$

$$\sum M_0 = (M + dM) - M - (Q + dQ)\left(\frac{Rd\varphi}{2}\right) - Q\left(\frac{Rd\varphi}{2}\right) + dF\mu r_1. \tag{44}$$

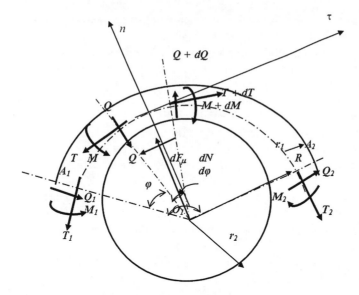

Figure 16. Friction in non-flexible fibres or yarns passing around a peg.

The analytical solution of the equilibrium equation in the closed form was given by

$$T_2 = T_1 e^{\mu K \theta} + f_0, \tag{45}$$

where the term f_0 given for fibres and yarns exhibits linear bending behaviour, i.e.

$$f_0 = \frac{\mu K \sqrt{B}}{R} \left(\sqrt{T_1} + \sqrt{T_2} \right), \tag{46}$$

and for yarns with non-linear bending behaviour,

$$f_0 = \frac{2m_0}{R} + \frac{\mu K \sqrt{B}}{R} \left(\sqrt{T_1} + \sqrt{T_2} \right), \tag{47}$$

where K is the curvature and B is the flexural rigidity of the yarn. As seen in the final equation, even the improved Capstan equation is limited by itself. The output tension cannot be obtained from the equation directly. The equation proposed by Wei and Chen [128] can be used to analyse the frictional coefficient of yarn or fibre by using Capstan method. However, as seen, it depends on the bending rigidity of the yarn or fibre, which is also difficult to assess.

As stated earlier, if incoming tension T_1 is zero, then the output tension is also zero. This is the limitation of the Capstan equation. Wei and Chen [128] claimed that by using their improved equation, the limitation of the Capstan equation would be overcome. They said '*when $T_1 = 0$, we have $T_2 = f_0 \neq 0$. This result not only helps to overcome the limitation of the Capstan equation (i.e.: that T_2 is always zero) but it can also be used to calculate the actual tension of fibres in spun yarn*'. If one examines the equation given for f_0, it can be seen that the conclusion given is not complete. First, f_0 is a function of T_2, which indicates that in any experiment the errors are correlated. It is possible to calculate the tension of the yarn from the formula given if fibres and yarns have a linear bending behaviour. However, for those with non-linear bending behaviour (and most of fibres and yarns fall into this category), it cannot be calculated directly because tension is a function in itself. Furthermore, the model proposed is valid for only small wrap angle and applicable to relatively rigid fibres and yarns. In common with most other investigators, Wei and Chen [128] also neglected the effect of the twist blockage and inertia moment of the element in their analysis.

During the second half of the twentieth century, interest in the frictional properties of fibres has increased. Since frictional properties of fibres are primarily responsible for the mutual adherence of fibres during processing, many investigators have proposed a friction model for fibrous materials. Most of the models do not include bending, viscoelastic properties and speed of the element into the model. Kowalski's frictional model [127] has included most of these terms, but the model is very complex and has almost no practical use. It was reported that there was a reasonable agreement between the adhesion theory of metal friction and polymer friction. Therefore, common concepts of the adhesion theory of metallic friction have been extended to include polymer friction. Table 3 summarises the friction models proposed by different investigators.

5. Measuring fibre friction

Over the past years, different types of instruments have been developed to measure the frictional characteristics of fibres. Some of the methods were used for fundamental research works, although others were used for practical investigations related to the processing. As has been mentioned earlier, the coefficient of friction is material dependent. Therefore, it is important to perform the friction test with different surfaces and report the values of the coefficient of friction according to the materials involved to the testing. Inter-fibre friction may have the first priority for assessing the frictional characteristics of fibres, and the second will be the fibre-to-metal friction [129].

5.1. Measurement methods

Although different types of instruments have been developed for the inter-fibre friction test, all of them exist in one of the following categories (Figure 17) [20,22,23,130,131]:

- Area-contact method.
- Line-contact method.
- Point-contact method.

Each of these methods is actually a simulation of the rubbing action. In textile processing, examples of procedures and deformational models that involve these three types of contacts are as follows:

- *Point-contact*: Opening and cleaning, carding, weaving and knitting.
- *Line-contact*: Drafting, needle punching, roving, spinning and weaving.
- *Area-contact*: Opening and cleaning, roving, drafting and spinning.

The last two methods usually require lengths of fibre that are long enough to permit sliding during the friction test as well as easy handling of the fibres. These methods also require a very large sample size in order to obtain a reliable result from the test. The point-contact and the line-contact methods can be used with synthetic filaments and long staple fibres such as wool and silk. However, they are not an adequate test method for

Table 3. Friction models proposed by different investigators.

Investigator	Proposed model	Year
Da Vinci, Amontons	$F = \mu N$	1690
Euler	$T_2 = T_1 e^{\mu\theta}$	—
Whitehead, McFalane, Tabor	$F = \mu(N)N$	1950
Olofsson	$F = \mu_0 N + \alpha A, \ T_2 = T_1 e^{\mu_0\theta} + \beta \dfrac{e^{\mu_0\theta} - 1}{\mu_0}$	1950
Bowden, Young	$F = \mu(N)N$	1951
Gralen, Makinson	$F = \alpha N + bN^{1-\beta}$	1952
Howell	$F = \alpha N^m$	1953
Howell, Lincoln	$T_2 = T_1 \left[1 + (1 - n)a \left(\dfrac{\rho}{T_1} \right)^{1-n} \alpha \right]^{\frac{1}{1-n}}$	1953
Garbaruk	$T_2 = T_1 e^{\mu\theta} + \dfrac{v^2}{Nmg}(e^{\mu\theta} - 1)$	1963
Fieles-Kahl, Helli	$T_2 = T_1 e^{\mu\theta} + \dfrac{R}{\mu}(e^{\mu\theta} - 1), \ R = \eta s \dfrac{dv}{dz}$	1966
Rogoza	$T_2 - T_1 e^{\mu\theta} + \dfrac{EJ}{2\rho^2}(e^{\mu\theta} - 1)$	1967
Kowalski	$T_2 = T_1 e^{\mu\theta} + \dfrac{ED^2}{12\rho}(e^{\mu\theta} - 1)$	1973
El Mogahzy	$F = SC_M \left(\dfrac{1}{K} \right)^n m^{1-n} N^n$	1987
Kowalski	$T^2 = H$ $+ \dfrac{nvT_1}{c\rho\alpha}\left\{ \left[1 - (1 - n)a \left(\dfrac{\rho}{T_1} \right)^{1-n} \alpha \right]^{\frac{1}{1-n}} - 1 \right\}$ $\left(1 - e^{-\frac{\rho dc_1}{v\eta}} \right)$	1991
El Mogahzy, Broughton	$F = a_0(nf)^{2(1-n)}$ and $a_0 = CSC_m K^{-n}\delta^{1-n}$	1993
Wei, Chen	$T_2 = T_1 e^{\mu K\theta} + f_0,$ for linear bending, $f_0 = \dfrac{\mu K \sqrt{B}}{R}(\sqrt{T_1} + \sqrt{T_2}),$ for non-linear bending, $f_0 = \dfrac{2m_0}{R} + \dfrac{\mu K \sqrt{B}}{R}\left(\sqrt{T_1} + \sqrt{T_2} \right)$	1998

cotton fibres. However, the area-contact method would be a better choice for assessing the frictional characteristics of cotton fibres. This is because, in practice, cotton fibres tend to move in groups rather than individually. The area-contact method can also give results that are more accurate in case simulation of the frictional behaviour of staple fibres [23,132–134].

5.1.1. Requirements for friction measurements

Regardless of the method being used, the measurement of the frictional characteristics of fibres should exhibit the following criteria, otherwise the test would be inaccurate.

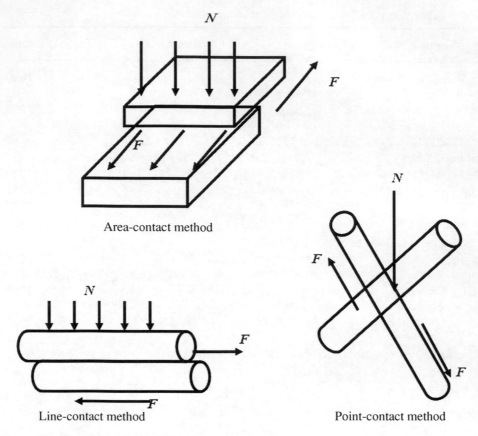

Area-contact method

Line-contact method

Point-contact method

Figure 17. Common methods of friction test.

- The method should be an actual simulation of a situation that occurs in the processing or during the deformation of fibre assembly [133,134].
- The method must allow rapid preparation of samples and rapid testing without violation of the accuracy criterion and need of operator skills. The instrument should be automated, or at least semi-automated, in order to eliminate the interaction with an operator. This criterion may be the most challenging task in developing a friction-measuring instrument [23,98,133].
- The data obtained from the instrument by using any one of the above-mentioned methods should be meaningful, accurate and, more importantly, reproducible, which is another challenging issue for the fibre friction evaluations [21,133,134].

The reproducibility of the data may be increased by taking some precautions before and during the testing. For example, the sampling of fibres must be taken carefully, and it should represent the entire population. Table 1 shows the static and kinetic coefficients of friction for cotton fibres from different investigators. As seen in the table, the range of friction seems to be highly spread out. This confirms that the reproducibility of a single-fibre friction test was not achieved. The reproducibility of the data could be achieved by using a large sample size and attributed to the following reasons:

- *Inconsistency of the test performance*: Handling of fibres requires experience and special skills. Handling of fibres is a major inconsistency of the friction testing. By using an automated instrument, this problem could become insignificant [20,98,135]. Data given in Table 1 obviously has a performance problem due to the inconsistency of the test.
- *Testing condition*: Traditionally, the coefficient of friction has been the main parameter measured by any friction-testing methods. Since textile fibres are more sensitive to the environmental conditions, a slight change in testing conditions will result in a wide variation of the coefficient of friction. Therefore, it is important to have the coefficient of friction determined by a test that is least influenced by the testing conditions [20,21,133].
- *Sensitivity of the test*: In the measurement of the fibre frictions, the normal load is in the order of mN, which means the friction test is very sensitive. In fact, it was reported that the friction test was more sensitive to the environmental conditions, such as relative humidity and temperature, than the strength test of the fibres. Therefore, prior to the test, surfaces must be free of environmental dust and foreign matter. The instrument should also be in a container so that the effect of air circulation could be eliminated. In case of testing fibres with high regain, special care should be taken for relative humidity [23].
- *Variability of fibres*: In the friction test, the characteristics of a fibre surface are being assessed. The fibre surface is identical from one specimen to another; however, fibre friction is correlated to several other fibre characteristics. Thus, the variability in the surface will be a complex function. The confidence interval for the fibre friction will be large due to high variability coming from other characteristics of fibres [20,98].

5.1.2. Measurements of fibre friction

The progress of measurement apparatus for fibre friction has been a major activity in many experimental studies [136]. Although much work has already been done on inter-fibre friction measurements, no standard apparatus has been developed. There are some basic features that are common in most of the methods designed for measuring friction between single fibres. It was reported that the apparatus developed by Bowden (see Figure 4) was the best general method for the fundamental studies of friction [23]. This apparatus was suitable for the loads between 5 mN and 100 cN. It is, however, very difficult to use the instrument for testing of fibrous materials. In inter-fibre friction testing, measurements are made using a pair of fibres, which have been identically treated and made to move relative to each other. In practice, the measurement of fibre friction can be made in three different methods, as given below:

- *Classical capstan method*: A fibre is passed over a cylindrical rod and the tension required to cause slippage is measured.
- *Classical law of friction*: The force of sliding one fibre against the normal pressure exerted by another fibre is measured.
- *Gralen and Lindberg* [50,93] *methods*: Two fibres are twisted together and the force required to cause the slippage is measured.

Gralen and his co-worker [93] conducted the above-described test by holding the fibres in contact at right angles, and putting an increasing torque on one fibre by means of a torsion wire. The fibres remained together until the torque was large enough to overcome the friction between them, at which one of the fibres slipped. In order to get an accurate

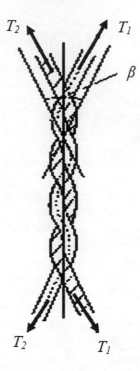

Figure 18. Twist-method diagram.

result from the apparatus developed by Gralen and Olofsson [93], special care should be taken when handling both the fibre specimen and instrument. Precise recording of the motion was also necessary for accurate results. In the experiment, the authors used a camera to record the movement of one fibre. Lindberg and Gralen [50,137] modified the method proposed by Gralen and Olofsson [93] and came up with another method known as the 'fibre-twist method'. In this method, the coefficient of friction between single fibres was measured by the insertion of a certain number of turns, and known weights were attached to one end of each fibre as shown in Figure 18. Then, the tension on the other ends of fibres was increased until the critical value at which the difference in tension was not supported by the friction between fibres. At the value where slippage occurs, the coefficient of friction can be calculated by using the following relationship:

$$\mu = \frac{\ln T_2 - \ln T_1}{\pi n \beta},$$
(48)

where T_1 and T_2 are tensions on the ends, β is the twist angle and n is the number of turns. The method is also suitable for the measurement of yarn friction. Howell [80] used another method in which one fibre was inclined at a variable angle, and the other was allowed to slide down it. The slip of fibres was detected by eye, and the angle of incline was read from a circular scale. Howell and Mazur [33] used a slightly modified apparatus to study the frictional characteristics of cellulose acetate and rayon fibres. They reported that although no abrasion on the specimens was seen, there was the possibility that some amount of wear might have modified the frictional properties of the fibres. Scheier and his co-worker [138] also used the same method in their experiment; however, they used an electronic force transducer system to convert the acting frictional forces to voltage output that was recorded

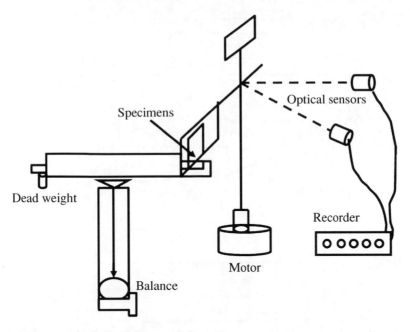

Figure 19. Schematic of Betts' friction apparatus.

on a chart. Although the method used by Scheier and Lyons [138] was more consistent, they failed to give the response time of the transducer used in the experiment. Betts et al. [139] developed, with some modifications, another inter-fibre friction apparatus based on the descriptions given by Gralen and Olofsson [93]. The schematic illustration of the apparatus is given in Figure 19. In the experiment, two identically treated fibres are placed into the jaws. One of the jaws is on the top of the balance that controls the normal weight, and the other is on the top of the helix screw that applies torque to the clamp. The movement of the fibre is detected by an optical system, which was capable of working up to 200 mg.

The line contact (fibre-twist method) test was widely used in the study of friction. Fair [140] and Fair and Gupta [141] studied the frictional characteristics of human hair. In their study, they developed an apparatus based on the method given by Gralen and Lindberg [50] to measure friction with a standard tension metre, as shown in Figure 20. The device can be clamped into the lower jaw of the Instron® machine. Specimens with the desired number of twist insertion are placed into the apparatus, as seen in the figure. Equal weights are tied to one fibre end, while the other ends are clamped to the load cell. Then the Instron® crosshead begins to move downward, carrying the device and the weights attached to the specimens. The tension builds up in the fibres until it overcomes the force of friction where slippage between the fibres takes place. The stick-slip pattern can be seen in the Instron® chart, from which the static and kinetic friction coefficients can be calculated. As seen in the figure, the device is not an appropriate choice for short fibres and can only be used with long, natural or synthetic fibres. The device is capable of making measurements in a liquid medium rather than air. Simply, the container can be filled with the desired liquid medium before testing. Table 4 shows the value of the coefficient of friction for fibres with the test methods and parameters.

Pascoe and Tabor described another apparatus used to measure the fibre friction. In the apparatus, the sliding fibre is placed at one end, and the other end rests on the second fibre, as shown in Figure 21. The vertical deflection due to cantilever action gave the value of

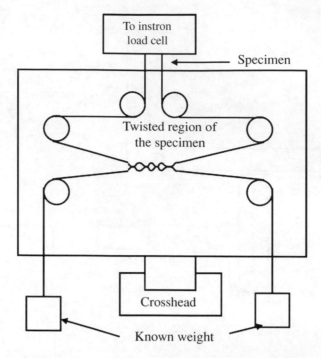

Figure 20. Schematic of the line-contact friction-testing device.

the normal force, and the horizontal deflection during the slippage of the fibre gave the frictional force. The deflection was observed with a microscope through a glass window mounted on the surrounding enclosure [20,23]. By using this instrument, Pascoe and Tabor were able to conduct their experiment in a load range of 1 μN to 1 cN normal force. The static and kinetic coefficients of friction can also be measured by the sled test method [133]. The test measures the force required to pull a known weight with a horizontal surface across a parallel sample of the substrate under the sled as seen in Figure 22. The method is applicable for fibre-to-fibre as well as fibre-to-metal friction tests.

Lord [142] came out with a different test method in which two sets of fibre fringes were used in the experiment. His method was based on sliding one fibre fringe against another under a lateral load by placing a rectangular block on the top of the contacting fibre fringe. He indirectly measured the frictional force by using a cantilever spring attached to the top fringe. El Mogahzy and Broughton [21] employed the method described by Lord [142] and developed an apparatus for the friction measurement of cotton fibres. As shown in Figure 23, a fibre beard, similar to that prepared for a fibrograph, is held in a clamp and a fixed surface area is rubbed over the fibre beard from top to bottom. If the applied normal force is kept constant, the resistance to sliding will progressively decrease as the sliding surface moves from the densest region to the region involving fewer fibres in contact. This decrease is due to the decrease in the number of fibres as the surface moves towards the ends of the longer fibres. If the friction is measured at different points of sliding, the result will be a relationship between the frictional force and the number of fibres contacting the fixed area surface. This relationship is commonly termed as 'the friction profile of fibres'. Figure 24 represents a typical friction profile of fibres. As seen in the figure, there is a relationship between the number of fibres and the friction force. For the fibre-to-metal friction test, the

Table 4. Values of friction of coefficient for fibres.

| Investigator | Measuring method | Test conditions | | | | Fibre type | μ_s |
		N (mg)	T_o (mg)	β	n		
Mercer	Point contact	30	—	—	—	Nylon 66	0.26
Mercer	Point contact	30	—	—	—	V. rayon	0.30
Mercer	Point contact	30	—	—	—	Acetate	0.39
Mercer	Point contact	30	—	—	—	Cuprammo	0.32
Gralen	Point contact	17	—	—	—	V. rayon	0.30
Olofsson	Point contact	17–97	—	—	—	Wool	0.61
Olofsson	Point contact	17–97	—	—	—	V. rayon	0.35
Olofsson	Point contact	17–97	—	—	—	Nylon 66	0.47
Hood	Twist method	—	400	6	6.25	Cotton	0.17
Hood	Twist method	—	400	6	6.25	Nylon	0.43
Hood	Twist method	—	400	6	6.25	V. rayon	0.46
Hood	Twist method	—	400	6	6.25	Cuprammo	0.30
Hood	Twist method	—	400	6	6.25	Ramie	0.49
Gralen	Twist method	—	2180	5.74	4.5	Wool	0.35
Gralen	Twist method	30	400–3000	5	4.5	Wool	0.11
Gralen	Twist method	30	400–3000	5	4.5	Cotton	0.22
Gralen	Twist method	—	—	—	—	Jute	0.46
Gralen	Twist method	—	—	—	—	Silk	0.52
Gralen	Twist method	—	—	—	—	Acetate	0.56
Gralen	Twist method	—	—	—	—	Terylene	0.58
Roder	Capstan method	—	0–1 g/cm	π	—	Wool-glass	0.60
Roder	Capstan method	—	0–1 g/cm	π	—	Wool-sheep's horn	0.63
Duckle	Capstan method	—	25 g	$<\pi$	—	Rayon steel	0.38
Duckle	Capstan method	—	25 g	$<\pi$	—	Acetate steel	0.38
Duckle	Capstan method	—	25 g	$<\pi$	—	Nylon steel	0.32
Duckle	Capstan method	—	25 g	$<\pi$	—	Cotton steel	0.29

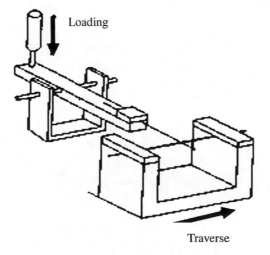

Figure 21. Pascoe and Tabor's fibre friction apparatus.

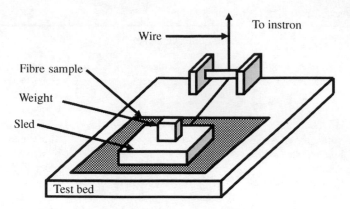

Figure 22.   Schematic of sled test apparatus.

sliding surface is just a metal plate, whereas, for the fibre-to-fibre friction test, the surface
is covered by fibres.

5.1.3. Measurement of yarn friction

Yarn friction has an important role in sewing, knitting and weaving. Machine parts are
always in contact with yarns; hence, it is an important issue to assess the frictional characteristics of yarns [76,143–150]. In the previous section, the measurement methods for the
fibre friction have been given. Theoretically, all methods used for fibres can also be used
for the yarn friction test, although some of them are not practical for the yarn friction test.
For that reason, other types of apparatus have been developed for the yarn friction test.
Furthermore, the methods given for fibre friction are suitable for the fundamental investigations and not for the rapid technical evaluations. The basic apparatus to measure the yarn
friction is the Capstan method. If the coefficient of friction is constant, the basic equation
$(T_2 = T_1 e^{\mu\theta})$ may be used to calculate the friction coefficient μ. The Capstan method can
be static or dynamic, as shown in Figure 25. The static Capstan method is suitable for
the determination of the coefficient of the static friction; conversely, the dynamic Capstan
method is used for assessing the coefficient of the kinetic friction. Buckle and Pollitt [151]
modified the dynamic Capstan method by attaching a scale so that the coefficient of friction

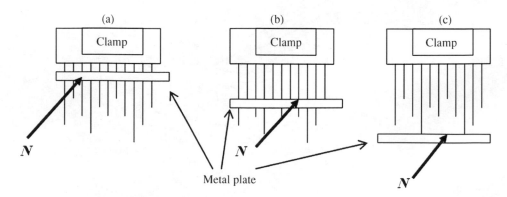

Figure 23.   Schematic representation of fibre beard test.

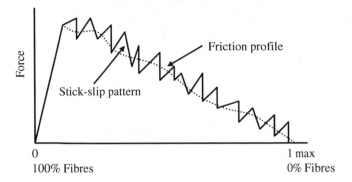

Figure 24. Friction profile of fibre.

can be read directly. Figure 26 shows a schematic representation of the apparatus developed by Buckle and Pollitt [151]. The calibration of the apparatus plays an important role in the measurement; therefore, calibrating charts based on the yarn type should be used.

The methods given above are primarily used for a yarn-to-metal friction test. For yarn-to-yarn, the methods should be slightly modified. The static Capstan method can be used for yarn-to-yarn friction test by covering the surface of the cylinder with the same yarn. Another method is the twisted strand method, as shown schematically in Figure 27. The American Society for Testing and Materials (ASTM) has two standard test methods for yarn-to-yarn and yarn-to-metal testings. Exact procedure and dimension of the apparatus have been given in the ASTM standards [152,153].

5.2. *Factors affecting friction of fibrous materials*

In the previous sections, several experimental techniques and theoretical models proposed for the measurement of frictional characteristics of fibres were given. As mentioned earlier, the friction test is more sensitive than most of the physical tests; therefore, it would be necessary to consider the factors affecting the frictional characteristics of fibres. For example,

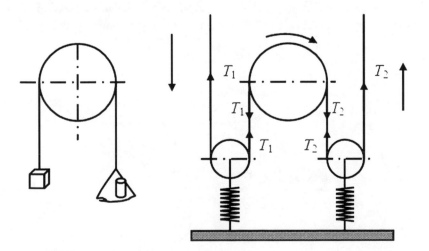

Figure 25. Capstan method.

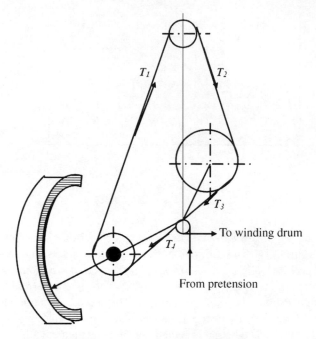

Figure 26.   Apparatus for automatic indication of coefficient of friction.

if a finish were applied to fibres, the general equations could not be applied to calculate the coefficient of friction. Olsen [123] has studied the frictional behaviour of a spin finish lubricated yarn. The relationship between the coefficients of friction, the yarn speed, lubricant viscosity and the applied pressure is in contact is given in Figure 28. The study of Olsen [123] was later confirmed by Schick [84,154–158]. Figure 28 shows the idealised general frictional behaviour of spin lubricated yarn. As seen in the figure, the total coefficient of friction is composed of boundary and hydrodynamic friction, and the minimum coefficient of friction is obtained at the semi-boundary region. The speed below 0.1 m/min is the boundary region and above 5 m/min is hydrodynamic region. At the boundary region, the mechanical, chemical and physical forces are dominant, although at the hydrodynamic

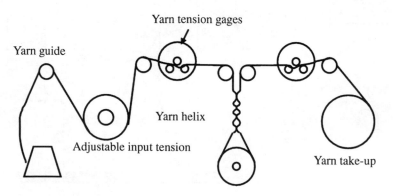

Figure 27.   Twisted yarn method.

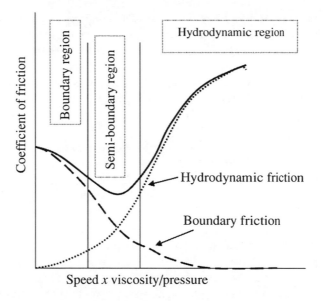

Figure 28. General frictional behaviour of liquid-lubricated yarn.

region, the characteristics of fiction depend upon the chemical structure, wetting and shear properties.

Studying the frictional behaviour of fibres in the yarn over a cylindrical surface is the most widely employed technique. In the process, the spin finish forms a film between the fibre and friction surface and the friction is dependent upon the shear strength of that film [123,156]. Lyne [122] reported that as the viscosity of the lubricant is increased, the final tension also gets increased. Figure 29 shows the effect of viscosity on friction for acetate fibres. Hansen and his co-worker [44] conducted experiments in which the yarn speed

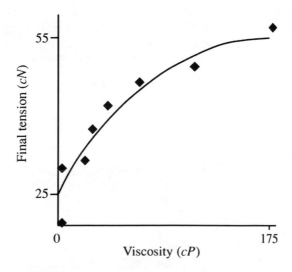

Figure 29. Friction of acetate yarn vs. lubricant viscosity.

was kept constant with variable viscosity. Their data confirmed Lyne's findings. Hansen and Tabor [44] also extended their experiment by studying the frictional force at constant viscosity with variable speed. They found that the plot of the final tension versus speed x viscosity was almost identical. Hence, they concluded that the friction depended only on the product of speed and viscosity. The shape of their plot is similar to the plot given by Lyne [122] in Figure 29. Furthermore, the following aspects can be given briefly as factors affecting the friction:

- *Effect of speed*: Schick [159] evaluated the friction dependency on the yarn speed with three different viscosities. He concluded that there was no difference in friction at low-speed ranges, despite the difference in viscosity. In contrast, at high-speed ranges, friction was dependent on viscosity. Röder [160,161] reported that the static coefficient of friction increases with increase in speeds ranging from 3 cm/min to 2 m/min. Park and his co-workers [162] studied the effect of speeds ranging from 1 cm/min to 400 m/min using polyester, polyamide and polypropylene fibres. In general, they observed a maximum in frictional force at a certain speed, above which frictional forces tended to decrease as the sliding speed is further increased [163].
- *Effect of temperature*: Mercer and Makinson [53] noticed the effect of temperature when studying the frictional characteristics of wool fibres. The effect of temperature is important because in processing, high spinning and drawing speeds result in temperature increase in fibres passing over guides. Schick [154] studied the effect of guide temperature and found that an increase in the guide temperature resulted in a decrease in friction. However, his findings about temperature dependency were also related to the viscosity of the lubricant because as temperature increases, the viscosity of the lubricant decreases.
- *Effect of denier*: Martin and Mittelmann [112] reported that there was a positive correlation between the frictional force of wool and fibre diameter. The frictional force was found to increase with an increase in the linear density of fibre. Schick [154] also found that friction increases with increasing the denier. Although the area of contact was not directly measured, he interpreted this increase as being due to an increase in the area of contact with large denier fibres [164].
- *Effect of contact angle*: The effect of contact angle that a yarn or fibre makes around a cylindrical pin has been studied, and it was found that the change of contact angle causes a proportional change in the contact area; within the boundary region, an increase in contact area led to an increase in fibre friction [84,154]. In the hydrodynamic region, the frictional force is given by

$$F = \eta \left(A \frac{V}{d} \right),   \tag{49}$$

where η is the viscosity of lubricant, A is the contact area, V is the yarn speed and d is the thickness of the lubricant film.
- *Effect of moisture regain*: It was found that the effect of moisture regain was dependent on the type of fibre being tested. In hydrophilic fibres, Schick [154] observed a marked increase in friction in the hydrodynamic region by increasing the relative humidity from 12 to 69%. This increase in friction was attributed to an increase in the area of contact between fibre and surface. Röder [160] found that the coefficient of friction of wool increased even further for relative humidity of over 70% [10,154,165,166].

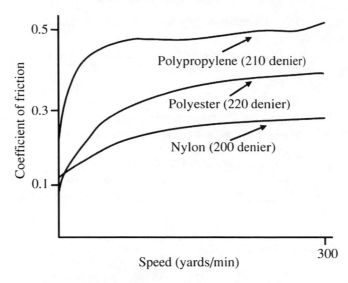

Figure 30. Effect of fibre material on friction.

- *Effect of guide surface roughness*: Olsen [123] reported that an increase in the roughness of a guide surface resulted in a decrease in friction. Schick [156] agreed with Olsen on this concept. He found that increasing the surface roughness of a chrome pin resulted in a decrease of tension at a speed of 50 m/min from 1.35 N to 0.80 N. Since this speed was in the hydrodynamic region, he interpreted it as a decrease in friction. However, at speeds below 10 m/min, the opposite phenomenon occurred, yet he failed to give an explanation for it [167].
- *Effect of fibre material*: It is now a well-established fact that the same lubricant can give results of different frictional forces with varying fibre types [168]. Figure 30 shows the coefficient of friction obtained from different fibre types with the same lubricant applied. This phenomenon is related to the wetting properties of fibres. Schick [156] stated that the poor wetting of polymer resulted in a higher coefficient of friction.
- *Effect of lubricant viscosity*: Lyne [122] found that at high speeds, friction increased with an increase in viscosity. Schick's experiment also showed that under a condition of good wetting, the friction was independent of the lubricant viscosity in the boundary region, but dependent in the hydrodynamic region [156,169–172].
- *Effect of molecular groups on fibre friction*: Röder [160] investigated the effect of molecular groups on fibre friction. He observed that higher numbers of ethylene oxide groups resulted in an increase in friction due to an increase in the viscosity component. It was also noted that there was very little difference between the effects of –COOH and –OH groups in the molecular structure on fibre friction; however, there was a decrease when an $-NH_2$ group was introduced into the structure [173]. Schick [155] agreed with Röder on this phenomenon, as seen in Figure 30.
- *Effect of cross section of fibres on friction*: 'Experiments performed on viscose tire cord indicated that the friction coefficients were not related to the shape of the cross section. These differences in cross section were caused only by varying manufacturing parameters or raw materials and not by changes in spinneret hole design' [96]. However, El Mogahzy [20] reported that the contact area was related to the cross

section of the fibre and found that the values of μ for non-circular fibres were lower than those of circular fibres.

- *Effect of the amount of lubricant on fibre*: The amount of lubricant on the fibre has been investigated in many studies [123,155,156,160,174,175]. Röder [160] found that 0.20% finish on the fibre surface resulted in a minimum value of friction. Olsen [123] agreed with Röder and found that a plot of minimum finish level versus friction at 0.15% fibre weight was attributed to the amount required for forming a mono-molecular film on the fibre. Schick [156] performed another group of experiments in which the amount of finish was varied from 0.25 to 2.0%. He reported that in the measurement of friction versus speed, the curves levelled off to a plateau at about 0.5% weight of fibre. Röder and Olsen did not consider the effect of speed on their experiments, and Schick did not run his experiment at below 0.5% finish. Therefore, the disagreement between Schick and the previous two authors exists due to the lack of concordance in the experimental design.
- *Effect of crystallinity on friction*: It was reported that as the crystallinity of nylon increased, the density, stiffness, yield stress, hardness and abrasion resistance also increased [176]. At the low levels of crystallinity, an increase in crystallinity resulted in a decrease in the dry coefficient of friction.
- *Effect of fibre lustre on friction*: This is an important surface property for synthetic fibres, and its intensity depends on the amount of TiO_2 added during polymerisation [177,178]. Generally, three types of fibre lustre are encountered in practice, which are as follows:

  - *Bright fibres*: They may contain up to 0.1% of TiO_2.
  - *Semi-dull fibres*: They contain up to 0.5% of TiO_2.
  - *Dull fibres*: They contain about 2.0% of TiO_2.

Scheier and Lyons [179] stated that variations in TiO_2 content had an influence on the surface geometry of fibres. Bright fibres have smooth surfaces, and dull fibres have rough surfaces. Schick [180] related the differences in fibre lustre to their frictional properties. He showed that, as the amount of lustre increased, the frictional forces also increased in the boundary and semi-boundary regions; however, in the hydrodynamic regions, the relationship was in the reverse order.

6. Compression of fibrous materials

6.1. Introduction

The main characteristics of fibrous material assembly should be soft, permeable to air, and compressible. During compression of fibre assembly, pressure P is applied to a fibre mass, and this pressure causes a volume change in the fibre assembly. Other physical changes occur during compression; however, the bending and surface properties of a fibre have the maximum influence on the compressive characteristics of fibres. The conventional method of conducting compression tests on fibrous masses consists of a 'piston-in-cylinder' arrangement. The energy required to compress fibre masses includes the energy expended in overcoming friction between fibres and cylinder wall, bending energy of fibres and crimp. The relationships between fibre properties and bulk compression characteristics of fibres have been studied in many experiments [35,87,181,182]. The study of compression of fibrous materials can be divided into two main groups, that is, the experimental techniques and theory.

6.2. *Theories on compression of fibrous materials*

Most of the theories, as well as the experiments developed, about the compression characteristics of fibrous materials were conducted with wool fibres. Although it was realised
that an important attribute of a fibrous mass was its behaviour under the compression,
special attention was paid to the wool fibres where quality and suitability were concerned
[183–185]. There have been many studies in the past in which the compression properties
of fibrous masses and many proposed measures for characterising these properties have
been investigated. In his early work, van Wyk [185] referenced the Eggert's equation for
the compression of wool by

$$\left(\frac{v}{v_0}\right)^\gamma (p + p_0) = p_0, \tag{50}$$

where v_0 is the initial volume, p_0 is the latent pressure of the wool at zero applied pressure
and γ is the measure of pliability. van Wyk stated that the indexes, p_0 and γ, were extremely
sensitive to experimental error and dependent upon the initial volume. Therefore, the
reproducibility of the data is hardly possible. Since Eggert's model was not satisfactorily
applicable to the compressive characteristics of fibres, van Wyk [185] reduced the problem
to a simple bending of the fibres in the assembly. He considered an ideal rod supported
horizontally with a large number of points equally spaced at distances $2b$ apart, with equal
downward forces acting midway between the points and supports. For small deflections,
the deflection and the force are related by the following equation:

$$F = \frac{24IY}{b^2}y, \tag{51}$$

where I is the moment of inertia and Y is the Young's modulus of elasticity. Since in a fibre
assembly, the orientation of fibre is randomly distributed, he employed the mean distance
and mean number of contacts in the assembly. The element length b was assumed equal to
the mean distance between fibres. After completing the analytical derivations, the relation
between pressure and volume was given by

$$p = \frac{KYm^2}{\rho^2}\left(\frac{I}{v^2} - \frac{I}{v_0^2}\right), \tag{52}$$

where K is regarded as undermined constant, m is mass of fibres and ρ is the fibre density.

The model only included the bending properties of the material; none of the other
parameters related to the properties of fibres were included. The initial volume of the
material should be obtained by extrapolation, which introduces another error to the model.
As seen in Equation (52), van Wyk's equation is completely specified by two parameters, that
is, the constant KY and the initial volume v_0. Therefore, two parameters have to be specified
for the given model. Measurement of these two parameters is not easy because of several
factors, such as fibre slippage, that take place during compression. Thus, reproducibility
of sample conditions is a major problem. Although the model is not perfectly applicable
to the compressive characteristics of fibrous materials, it was the initiation of the studies
related to the compression of the fibrous materials. Therefore, it has a historical respect in
the area of studying compression behaviour of fibrous materials.

The number of contacts has a fundamental importance in the determination of mechanical properties of fibre assemblies. van Wyk [185] calculated the average distance between

neighbouring points on a fibre in a random assembly of uniform cylindrical fibres and gave the following formula for it:

$$\bar{l} = \frac{2V}{\pi DL}, \tag{53}$$

where V is the volume of the assembly, D is the diameter of the fibre and L is the total length of fibres. This result was derived on the assumption that the fibres are oriented randomly. Stearn [184] did not accept random orientation after compression and modified van Wyk's approach by taking into account the change in orientation of fibre in the assembly during compression. He gave the number of contacts per unit volume as

$$n_v = \frac{4}{\pi} L_c^2 D \frac{2}{\pi \sigma} \int_0^{\pi/2} I_c \xi \left(\varphi_c \right) d\varphi_c, \tag{54}$$

where L_c denotes the total length of fibres in a unit volume, σ is the compressibility, ϕ_c is the angle between the direction of compression and the centreline of the fibre. I_c and $\xi(\phi_c)$ are given as follows:

$$I_c = \int_0^{\pi/2} \sqrt{\sin^2 \varphi_c \cos^2 \gamma \left(1 - \sigma^2 \right) + \sigma^2 d\gamma}, \tag{55}$$

$$\xi(\varphi_c) = \sigma \sin \varphi_c \left(\sigma^2 \cos^2 \varphi_c + \sin^2 \varphi_c \right)^{\pi/2}. \tag{56}$$

The formulas given by Stearn are very complicated with restricted applications. Kallmes and his co-worker [186], on the other hand, derived the following expression for the number of fibre intersections in an area A:

$$v = \frac{N^2 \lambda^2}{\pi A}, \tag{57}$$

where N is the number of total fibres of length λ. It was assumed that all fibres lay in a plane with their directions perfectly random, and the thickness of the assembly was twice the diameter of these fibres. Although the expression given is simple, it cannot be applied if the fibres form a three-dimensional structure.

Komori and Makishima [187] derived a formula for the number of fibre intersections in a unit volume. They made the following assumptions:

- All fibres are straight cylinders of diameter D and length λ.
- The distribution of the centres of the mass of the fibre is random.
- Contacts of end-to-end or side-to-end are negligible.

Based on the above assumptions, they gave the following equation for the number of contact points in fibre assemblies with arbitrary distribution of orientation and fibre length:

$$\bar{n}_v = DL_v^2 I, \tag{58}$$

where I was given by

$$I = \int_0^\pi \int_0^\pi \int_0^\pi \int_0^\pi Y\left(\theta, \theta', \varphi, \varphi'\right) \Omega(\theta, \varphi)\Omega(\theta', \varphi')\sin\theta, \sin\theta' d\varphi' d\theta' d\varphi d\theta,$$

where

$$Y = \sqrt{1 - \{\cos\theta\cos\theta' + \sin\theta\sin\theta'\cos(\varphi - \varphi')\}^2}, \tag{59}$$

where Ω is the density function of orientation, θ, ϕ, θ' and ϕ' are respective polar angles characterising the orientation of two fibres to be brought in contact to each other. The model depends on the diameter D, the length L and the density function Ω. The measurement of diameter and length can be done easily for a given fibre assembly; however, there is no general method for the determination of density function. Therefore, the density function should be estimated for the fibrous mass. Komori and Makishima [187] considered two special cases for the density function, namely (1) assembly of random orientation in which the density function is $1/2\pi$, and (2) sheet-like assembly in which the density function is $(1/\pi)\delta(\theta - \pi/2)$. The first case confirms the van Wyk's equation [185] and the second one confirms the Kallmes and Corte's equation [186].

Komori and Makishima's analysis [187] was then taken as a starting point for most of the studies [188–190]. Itoh said that, for example, '*Hsu and Lee started from Komori and Makishima's analysis on treating the fibre contact in an isotropic masses, and explored a theory that makes it possible to relate the tangential elastic modulus and Poisson's ratio of an isotropic mass to orientation density and properties of the fibres*'. Carnaby and Pan [87] then extended the model given by Hsu and Lee, taking into account the effect of fibre slippage. Carnaby and Pan were able to reproduce the compression hysteresis of fibrous masses. In the analysis, fibres were divided into two categories as either slipping or non-slipping and analysed separately. The model given had a good agreement with the experiment at high compression ratios. They suggested that the contribution from the fibre viscoelasticity could be taken into account so that the model can be more accurate. Carnaby and Pan [87,191,192] developed a theory on the shear deformation of fibre masses based on their previous study [87]. The derivation of the theory is complex, yet it is valid for low-strain cases. Although the theories about the compression of fibre assemblies cast new light on the properties of fibres related to compression, the bending behaviour of fibres was not included directly in the models. The mean free fibre length was solely considered by averaging the distance between two adjacent fibres in contact over all of the mass (orientation independent). Komori and Itoh [189] extended the theories developed by Komori and Makishima [187] and Carnaby and Pan [87,192] by including the bending element of orientation-dependent mean length. The description of the model was based on the energy method. They claimed that the model satisfied more readily the theory of the compression mechanics of fibre masses. However, the density functions used in the model were not specified explicitly as in the model given by Komori and Makishima [187]. Although the notations used were similar to each other, they are totally different due to assumptions of the fibre orientations.

Abbott and Nason [188] considered the frictional characteristics of wool fibres when fibres are compressed. Before introducing the coefficient of friction into the analysis, when pressure is applied to a fibre sample, the pressure P at a distance x is given by

$$p = p_0 e^{-qx}. \tag{60}$$

The pressure decreases by a constant q, given by

$$q = 2\mu\alpha \left(\frac{a+b}{ab}\right), \tag{61}$$

where a and b are the respective height and width of the press box, and the parameter μ is the coefficient of friction and is assumed to be a constant. They derived an equation for the coefficient of friction in terms of loads and dimensions of the press box. The equation for the coefficient of friction was given as follows:

$$\mu = \frac{A_s}{2d(a+b)} \left(\frac{L_0}{L_s}\right) \left(\frac{L_d}{L_0}\right)^{s/d} \ln\left(\frac{L_0}{L_d}\right), \tag{62}$$

where d is the length of the fibre assembly after compression and Ls are the loads. The coefficient of friction is not a constant but a function of applied pressure. Then, an index called 'shear load gradient' was introduced for comparative purposes and is given by $\frac{L_0 - L_d}{d}$.

Although the model given is quite complicated in terms of parameters and their relationship, the shear load gradient is a good estimate of the coefficient of friction at various stages. The given experimental results showed a good correlation between the coefficient of friction and the shear load gradient index for wool, jute and high-density polyethylene fibres at different relative humidity and use of some type of petroleum grease and wax [188]. The bending properties of fibres were not considered in this model. Furthermore, Neckar [190] has developed a model for the mechanical behaviour of a fibrous material under one-dimensional and multi-dimensional compression due to normal stress. Although he did not include the effects of fibre surface properties into the model, he pointed out that the solution of the problem should be generalised to account for the influence of fibre-to-fibre friction and some other forms of energy dissipation, such as bending and crimp [190].

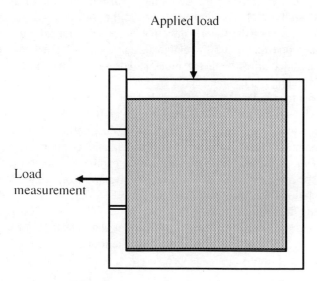

Figure 31. Schematic of Hearle's instrument.

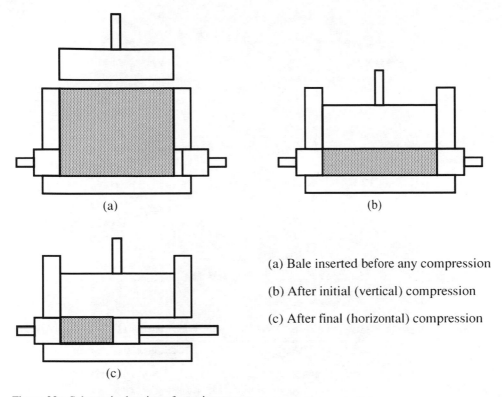

(a) Bale inserted before any compression

(b) After initial (vertical) compression

(c) After final (horizontal) compression

Figure 32.   Schematic drawing of pressing stages.

6.3. Experimental methods for compression of fibrous masses

The developed theories about the compression of fibrous assemblies have been confirmed by experimental studies with laboratory presses. However, in industry, the compression

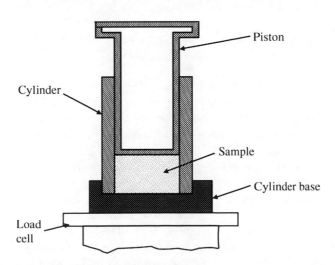

Figure 33.   Conventional piston-in-cylinder arrangement.

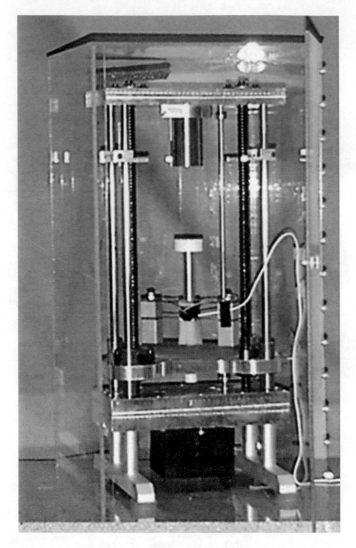

Figure 34.   Photograph of compression tester instrument.

presses are widely employed to package cotton and wool fibres. The compression of wool
has economical importance due to the cost of storage and transportation [178].

Hearle and El-Behery [193] developed an apparatus to measure transverse stresses
in fibre assemblies under loads. They placed the specimen in a rectangular container
and compressed by a load applied to the upper movable plate. Part of one side wall of
the container was also movable. A force-measuring device was mounted to the movable
side of the container in order to acquire the lateral pressure on the specimen, as seen in
Figure 31. They did not consider the frictional forces among fibres and between fibres and
the side walls of the container. Abbott and Nason [188], on the other hand, used a press
shown in Figure 32 to compress wool fibres. They placed a sensor on the instrument in
order to monitor the sample length during compression; however, they failed to provide
experimental techniques used during data acquisition as well as the response time for the
transducers.

A more realistic compression unit, used in many experiments, is a piston-in-cylinder compression cell [194,195,196]. Dunlop [194] employed the piston-in-cylinder apparatus to estimate the van Wyk's model parameters. He stated that there was considerable difficulty in obtaining the parameters required to characterise the compression properties of wool by means of conventional piston-in-cylinder apparatus. Young and Dricks [196] studied the energy required to compress wool fibres from the initial volume v_0 to the final volume v. They adapted the conventional method of conducting compression tests on fibrous masses by employing a 'piston-in-cylinder' arrangement, as shown in Figure 33. They performed the experiments and measured the energy required for the compression of samples from the load compression and relaxation curves. Although they claimed that boundary friction effects were eliminated by using pressure plate, the fibre-to-metal (cylinder wall) friction effect was still in the experiment. Therefore, the results are also overestimated due to the friction between the fibres and the cylinder wall. However, the method is potentially useful for estimating the compression characteristics of fibres. On the other hand, Yuksekkaya and Oxenham [197–202] proposed another type of compression-testing unit to measure the frictional properties of fibres. The instrument is a basic piston cylinder application with high sampling data acquisition connection. Figure 34 shows the picture of their proposed instrument. In their experiment, they have reported that it would be possible to distinguish the surface differences on fibre samples.

7. Conclusion

Unfortunately, the classical empirical friction laws do not hold true for fibrous and viscoelastic materials that constitute most of the textile fibres. In the second half of the twentieth century, fibre surfaces have been studied intensively for the frictional characteristics of different types of fibres. Most of the researchers have aimed to develop a test method and a test device that could be used to measure the frictional characteristics of fibres quickly, accurately and easily. Unfortunately, it was not possible to develop a standard test method or a test device for the measurement of textile fibres' friction. Most of the properties of cotton fibres can be assessed by using a HVI fibre-testing instrument. The HVI can process the fibre testing very fast. However, the frictional characteristics of a material (especially fibrous material) cannot be assessed rapidly, aside from there being so many variables involved in the nature of the problem.

The main objective of researches on the fibre friction area was to develop a system for giving an overall assessment of 'frictional characteristics' of fibres. In order to have a reliable and fast testing procedure, the methods should be rapid, reproducible, unbiased and not influenced by the operator. Due to the complexity of the fibre friction problem, it would be necessary to complete more research before totally understanding the friction properties of fibrous materials and to develop a fast and reliable testing apparatus.

References

[1] Encyclopaedia of Britannica, *Friction*, Vol. 5, Britannica Corporate, Chicago, IL, 1993.
[2] D. Halliday and R. Resnick, *Fundamentals of Physics*, 2nd ed., John Wiley & Sons, New York, NY, 1981.
[3] A. Harnoy, Tribol. Trans. 40 (1997), p. 360.
[4] W.R. Chang, I. Etsion, and D.B. Bogy, J. Tribol. 110 (1988), p. 57.
[5] T.C. Hsu and C.H. Lee, J. Manuf. Sci. Eng. 119 (1997), p. 563.
[6] A. Kyllingstad and G.W. Halsey, SPE Drill. Eng. 3 (1988), p. 369.
[7] A. Dutta, J. Appl. Mech. 50 (1983), p. 863.

[8] A.K. Chopra, *Dynamics of Structures: Theory and Applications to Earthquake Engineering*, Prentice Hall, Upper Saddle River, NJ, 1995.

[9] D.S. Larson and A. Fafitis, J. Eng. Mech. 121 (1995), p. 289.

[10] C.R. Wylie and L.C. Barrett, *Advanced Engineering Mathematics*, 6th ed., McGraw-Hill, New York, NY, 1995.

[11] E.H. Freitag, Contemp. Phys. 2 (1960), p. 198.

[12] E. Rabinowicz, Sci. Am. 194 (1956, May), p. 109.

[13] D.S. Taylor, J. Text. Inst. 46 (1955), p. P59.

[14] Ş. Ülkü, B. Özipek, and M. Acar, Text. Res. J. 65 (1995), p. 557.

[15] J.O. Ajayi, Text. Res. J. 62 (1992), p. 52.

[16] A.R. Kalyanaraman and R. Prakasam, Text. Res. J. 57 (1987), p. 307.

[17] V. Subramanian, G. Nalankilli, and V. Mathivanan, Text. Res. J. 57 (1987), p. 369.

[18] W.L. Balls, *Studies of Quality in Cotton*, Macmillan, London, 1928.

[19] G.-W. Du and J.W.S. Hearle, Text. Res. J. 61 (1991), p. 289.

[20] Y.E. El Mogahzy, *A study of the nature of friction in fibrous materials*, Ph.D. diss., North Carolina State University, Raleigh, 1987.

[21] Y.E. El Mogahzy and R.M. Broughton, Text, Res. J. 63 (1993), p. 465.

[22] H.G. Howell, K.W. Meiszkis, and D. Tabor, *Friction in Textiles*, The Textile Institute, London, 1959.

[23] W.E. Morton and J.W.S. Hearle, *The Physical Properties of Textile Fibres*, The Textile Institute, London, 1995.

[24] R.B. Finch, Text. Res. J. 21 (1951), p. 383.

[25] B.S. Gupta, *Friction in Textile Materials*, Woodhead Publishing Company, Cambridge, UK, 2008.

[26] Z.V. Shampai, Tech. Text. Ind. USSR 5 (1962), p. 151.

[27] W. Udomkichdecha, *On the compressional behavior of bulky fibre webs (Nonwovens)*, Ph.D. diss., North Carolina State University, Raleigh, 1986.

[28] F.P. Bowden and D. Tabor, *The Friction and Lubrication of Solids*, 2nd ed., Clarendon Press, Oxford, UK, 1954.

[29] F.P. Bowden and D. Tabor, *Friction: An Introduction to Tribology*, Anchor Press/Doubleday, Garden City, New York, 1973.

[30] I. Etsion and M. Amit, J. Tribol. 119 (1993), p. 406.

[31] G.B. Lubkin, Phys. Today 50 (1997), p. 17.

[32] M.E. Baird and K.W. Mieszkis, J. Text. Inst. 46 (1955), p. P101.

[33] H.G. Howell and J. Mazur, J. Text. Inst. 44 (1953), p. T59.

[34] B. Lincoln, J. Appl. Phys. 3 (1952), p. 260.

[35] D. Frishmen, A.L. Smith, and M. Harris, Text. Res. J. 18 (1948), p. 475.

[36] S.A.H. Ravandi, K. Toriumi, and Y. Matsumoto, Text. Res. J. 64 (1994), p. 224.

[37] L.B. Deluca and D.P. Thibodeaux, Text. Res. J. 62 (1992), p. 192.

[38] M.F. Belov, Tech. Text. Ind. USSR 4 (1964), p. 47.

[39] G. Brook and M. Hannah, J. Text. Inst. 46 (1955), p. P23.

[40] D.S. Taylor, J. Text. Inst. 48 (1957), p. T466.

[41] H.M. El Behery, Text. Res. J. 38 (1968), p. 321.

[42] R.S. Merkel, Text. Res. J. 33 (1963), p. 84.

[43] R.B. Finch, Text. Res. J. 21 (1951), p. 375.

[44] W.W. Hansen and D. Tabor, Text. Res. J. 27 (1957), p. 300.

[45] K. Harakawa, N. Oikawa, S. Takagi, and K. Tanaka, J. Text. Mach. Soc. Jpn. 23 (1977), p. 8.

[46] L.W. James, C.C. Cheng, and E.D. Kernite, Text. Res. J. 46 (1976), p. 496.

[47] A. Adderley, J. Text. Inst. 13 (1922), p. 249.

[48] S.B. Bandyopadhyay, Text. Res. J. 21 (1951), p. 659.

[49] S.C. Basu, A.A. Hamza, and F. Sikorski, J. Text. Inst. 69 (1978), p. 68.

[50] J. Lindberg and N. Gralen, Text. Res. J. 18 (1948), p. 287.

[51] K.R. Makinson, J. Text. Inst. 38 (1947), p. T332.

[52] K.R. Makinson and C. King, J. Text. Inst. 41 (1950), p. T493.

[53] E.H. Mercer and K.R. Makinson, J. Text. Inst. 38 (1947), p. T227.

[54] J.A. Morrow, J. Text. Inst. 29 (1938), p. T425.

[55] K.R. Sen and N. Ahmed, J. Text. Inst. 29 (1938), p. T258.

[56] K.N. Seshan, J. Text. Inst. 64 (1973), p. 631.

[57] K.N. Seshan and T. Svinivasan, J. Text. Inst. 64 (1973), p. 638.

[58] K.N. Seshan, J. Text. Inst. 66 (1975), p. 103.

[59] K.N. Seshan, J. Text. Inst. 66 (1975), p. 109.

[60] K.N. Seshan, J. Text. Inst. 69 (1978), p. 214.

[61] J.B. Speakman and E. Stott, J. Text. Inst. 22 (1931), p. T339.

[62] P. Grosberg, J. Text. Inst. 46 (1955), p. T233.

[63] P. Grosberg, J. Text. Inst. 54 (1963), p. T223.

[64] P. Grosberg and P.A. Smith, J. Text. Inst. 57 (1966), p. T15.

[65] S.L. Anderson, D.R. Cox, and L.D. Hardy, J. Text. Inst. 43 (1952), p. T362.

[66] M.A. Chaudri and K.L. Whiteley, J. Text. Inst. 60 (1969), p. 37.

[67] C. King, J. Text. Inst. 41 (1950), p. T135.

[68] K.W. Wolf, *An experimental study of interfiber friction in surgical sutures by the twist method*, M.Sc. thesis, North Carolina State University, Raleigh, 1979.

[69] M.V. Vspenskaya, Tech. Text. Ind. USSR 4 (1972), p. 20.

[70] B.G. Hood, Text. Res. J. 23 (1953), p. 495.

[71] R.K. Mody and P.V. Subramanian, Text. Dig. Sep (1971), p. 127.

[72] R.K. Mody and P.V. Subramanian, Text. Dig. Dec (1971), p. 177.

[73] W. F. du Bois and A. Viswanathan, J. Text. Inst. 57 (1966), p. T262.

[74] G. Quintelier, M. Warzee, and R. Sioncke, J. Text. Inst. 48 (1957), p. P26.

[75] A. Viswanathan, J. Text. Inst. 64 (1973), p. 553.

[76] H. Navkal and A. Turner, J. Text. Inst. 21 (1930), p. T511.

[77] B.A.K. Andrews, Text. Res. J. 59 (1989), p. 675.

[78] G.L.L. Louis, Text. Res. J. 57 (1987), p. 339.

[79] G. Lombard, O.G. Vermeersch, M. Weltrowski, J.Y.F. Drean, and C. Riss, Exper. Tech. 18 (1994), p. 9.

[80] H.G. Howell, J. Text. Inst. 42 (1951), p. T521.

[81] J.C. Guthrie and P.H. Oliver, J. Text. Inst. 43 (1952), p. T579.

[82] B. Lincoln, J. Text. Inst. 45 (1954), p. T92.

[83] M.M. Robins, R.W. Rennell, and R.D. Arnell, J. Phys. D Appl. Phys. 17 (1984), p. 1349.

[84] M.J. Schick, *Friction and Lubrication of Synthetic Fibres*, Marcel Dekker, New York, NY, 1975.

[85] N. Adams, J. Appl. Poly. Sci. 7 (1963), p. 2075.

[86] N. Adams, J. Appl. Poly. Sci. 7 (1963), p. 2105.

[87] G.H. Carnaby and N. Pan, Text. Res. J. 59 (1989), p. 275.

[88] F. Breazeale, Text. Res. J. 17 (1947), p. 27.

[89] J.A. Chapman, M.E. Pascoe, and D. Tabor, J. Text. Inst. 46 (1955), p. P3.

[90] K.L. Hertel and R. Lawson, Text. Bull. May (1968), p. 23.

[91] J. Lindberg, J. Text. Inst. 41 (1950), p. T331.

[92] H.G. Howell, J. Text. Inst. 45 (1954), p. T575.

[93] N. Gralen and B. Olofsson, Text. Res. J. 17 (1947), p. 488.

[94] N. Gralen, B. Olofsson, and J. Lindberg, Text. Res. J. 23 (1953), p. 623.

[95] M.R. Popovic and A.A. Goldenberg, IEEE Trans. Robotic Autom. 14 (1998, Feb.), p. 114.

[96] P.E. Slade, *Handbook of Fibre Finish Technology*, Marcel Dekker, New York, NY, 1998.

[97] K.R. Makinson and C. King, J. Text. Inst. 41 (1950), p. T407.

[98] B.S. Gupta and Y.E. El Mogahzy, Text. Res. J. 61 (1991), p. 547.

[99] Y.E. El Mogahzy and B.S. Gupta, Text. Res. J. 63 (1993), p. 219.

[100] Y.E. El-Mogahzy and R.M. Broughton, *Descriptive evaluation of fibre-machine interaction in spinning*, in *Beltwide Cotton Production Research Conference, Cotton Quality Measurement Session*, National Cotton Council, Memphis, TN, Jan. 1992.

[101] R.M. Broughton, Y.E. El Mogahzy, and D.M. Hall, Text. Res. J. 62 (1992), p. 131.

[102] T. Hsu and C. Lee, Tribol. Trans. 40 (1997), p. 367.

[103] B.J. Lance and F. Sadeghi, J. Tribol. 115 (1993), p. 445.

[104] A. Thompson and M.O. Robbins, Science 250 (1990, 9 Nov.), p. 792.

[105] C. Gao and D. Kuhlmann-Wilsdorf, J. Tribol. 112 (1990), p. 354.

[106] H. Yoshizawa, P. McGuiggan, and J. Israelachvili, Science 259 (1993, 26 Feb.), p. 1305.

[107] S. Jang and J. Tichy, J. Tribol. 119 (1997), p. 626.

[108] P. Grosberg and D.E.A. Plote, J. Text. Inst. 60 (1969), p. 268.

[109] P. Grosberg and D.E.A. Plote, J. Text. Inst. 62 (1971), p. 116.

[110] K.R. Makinson, J. Text. Inst. 61 (1970), p. 465.
[111] A. Viswanathan, J. Text. Inst. 57 (1966), p. T30.
[112] A.J.P. Martin and R. Mittelmann, J. Text. Inst. 37 (1946), p. T269.
[113] V.E. Gonsolves, Text. Res. J. 20 (1950), p. 28.
[114] B. Olofsson, Text. Res. J. 20 (1950), p. 476.
[115] H.G. Howell, J. Text. Inst. 44 (1953), p. T359.
[116] H.G. Howell, Text. Res. J. 23 (1953), p. 589.
[117] N.N. Matsuzawa and N. Kishii, J. Phys. Chem. 101 (1997), p. 10045.
[118] J. Mazur, J. Text. Inst. 46 (1955), p. T712.
[119] T. Nogia, Y. Narumi, and M. Ihara, J. Text. Mach. Soc. Jpn. 21 (1975), p. 41.
[120] T. Nogia, M. Ihara, and Y. Narumi, J. Text. Mach. Soc. Jpn. 24 (1978), p. 69.
[121] B. Olofsson and N. Gralen, Text. Res. J. 20 (1950), p. 467.
[122] D.G. Lyne, J. Text. Inst. 46 (1955), p. P112.
[123] J.S. Olsen, Text. Res. J. 39 (1969), p. 31.
[124] S. Galuszynski and R. Ellis, Text. Res. J. 53 (1983), p. 462.
[125] J.F. McMahou, J. Text. Inst. 79 (1988), p. 676.
[126] S. Kawabata and H. Morooka, J. Text. Mach. Soc. Jpn. 31 (1978), p. T96.
[127] K. Kowalski, Melliand Textlber. 72 (1991), p. E65.
[128] M. Wei and R. Chen, Text. Res. J. 68 (1998), p. 487.
[129] L. Virto and A. Naik, Text. Res. J. 67 (1997), p. 793.
[130] P. Ehler, Melliand Textilber. May (1977), p. 355.
[131] C.O. Grahan, P.L. Rhodes, and R.J. Harper, Text. Res. J. 47 (1977), p. 102.
[132] R.M. Broughton and Y.E. El-Mogahzy, *The measurement and importance of cotton fibre friction*, in *Beltwide Cotton Production Research Conference Cotton Quality Measurement Session*, National Cotton Council, Memphis, TN, Jan. 1992, pp. 1465–1467.
[133] R.M. Broughton and Y.E. El Mogahzy, National Cotton Council, Tappi. J. 76 (1993, Feb.), p. 178.
[134] Y.E. El-Mogahzy and R.M. Broughton, *Cotton fibres friction: Role, theory, and measurement*, in *Beltwide Cotton Production Research Conference, Cotton Quality Measurement Session*, National Cotton Council, San Antonio, TX, Jan. 1991, pp. 889–893.
[135] R.W. Ramirez, *The FFT Fundamentals and Concepts*, Prentice-Hall, Inc., Upper Saddle River, NJ, 1985.
[136] W. Bobeth, A. Mally, and M. Grun, Deutsche Textiltechnik 22, 6 (1972), p. 377.
[137] J. Lindberg and N. Gralen, Text. Res. J. 19 (1949), p. 183.
[138] S.C. Scheier, and W.J. Lyons, Text. Res. J. 35 (1965), p. 385.
[139] J.A. Betts, F.N. Hurt, K.D. O'Connor, and J. Repath, Hatra Res. Rep. 31 (1974, August).
[140] N.J.B. Fair, *The effect of chlorination of friction and surface morphology of dark brown and blond human hair fibres*, Ph.D. diss., North Carolina State University, Raleigh, 1984.
[141] N. Fair and B.S. Gupta, J. Soc. Cosm. Chem. 33 (1982), p. 229.
[142] E. Lord, J. Text. Inst. 46 (1955), p. P41.
[143] B.S. Gupta and P.T. Chang, Text. Res. J. 46 (1976), p. 90.
[144] M.J. Hammersly, J. Text. Inst. 64 (1973), p. 108.
[145] H.G. Jong, Text. Res. J. 63 (1993), p. 14.
[146] W.M. Koza, Text. Res. J. 45 (1975), p. 639.
[147] T. Nakashima and K. Ohta, J. Text. Mach. Soc. Jpn. 12 (1966), p. 185.
[148] D.G. Padfield, J. Text. Inst. 46 (1955), p. T71.
[149] M.W. Pascoe, J. Text. Inst. 50 (1959), p. T653.
[150] C. Rubenstein, J. Text. Inst. 49 (1958), p. T181.
[151] H. Buckle and J. Pollitt, J. Text. Inst. 39 (1948), p. T199.
[152] American Society for Testing and Materials (ASTM), *Standard test method for coefficient of friction, yarn to solid material, ASTM designation D3108-95*, in *ASTM Book of Standards*, Section 7 Vol. 7.01, ASTM, Philadelphia, PA, 1997, pp. 814–819.
[153] American Society for Testing and Materials (ASTM), *Standard test method for coefficient of friction, yarn to yarn, ASTM designation D3412-95*, in *ASTM Book of Standards*, Section 7 Vol. 7.02, ASTM, Philadelphia, PA, 1997, pp. 10–14.
[154] M.J. Schick, Text. Res. J. 43 (1973), p. 254.
[155] M.J. Schick, Text. Res. J. 43 (1973), p. 342.
[156] M.J. Schick, Text. Res. J. 43 (1973), p. 103.

[157] M.J. Schick, Text. Res. J. 50 (1980), p. 675.
[158] V.K. Srivastava, W.J. Onions, and P.P. Townend, J. Text. Inst. 67 (1976), p. 447.
[159] M.J. Schick, Text. Res. J. 43 (1973), p. 198.
[160] H.L. Röder, J. Text. Inst. 44 (1953), p. T247.
[161] H.L. Röder, J. Text. Inst. 46 (1955), p. P84.
[162] K. Park, C.G. Seefried, and G.M. Braynt, Text. Res. J. 44 (1974), p. 692.
[163] F. Konda, M. Okamura, A.M. Akbar, and T. Yokoi, Text. Res. J. 66 (1996), p. 343.
[164] P.K. Gupta, J. Am. Ceram. Soc. 74 (1991), p. 1692.
[165] P. Kenins, Text. Res. J. 64 (1994), p. 722.
[166] C. Wood, J. Text. Inst. 45 (1954), p. T794.
[167] E. Bradbury and A. Reicher, J. Text. Inst. 43 (1952), p. T350.
[168] L.J. Postle and J.I. Ingham, J. Text. Inst. 43 (1952), p. T77.
[169] N.V. Gitis and C. DellaCorte, Lubr. Eng. 51 (1995), p. 336.
[170] D. Hayes, Text. Month. July (1972), p. 38.
[171] D. Hayes, Text. Month. Aug (1972), p. 46.
[172] D. Wilson and M.J. Haminersley, J. Text. Inst. 57 (1966), p. T199.
[173] C. Wood, J. Text. Inst. 43 (1952), p. T339.
[174] A.M. Akbar, F. Konda, M. Okamura, and E. Maruh, Text. Res. J. 67 (1997), p. 643.
[175] C. Rubenstein, J. Text. Inst. 49 (1958), p. T13.
[176] R.T. Steinbuch, Mod. Plastics 42 (1964), p. 137.
[177] J. Lindberg, Text. Res. J. 18 (1948), p. 470.
[178] T.B. Sinha, Indian Text. J. 87 (1977), p. 121.
[179] S.C. Scheier and W. J. Lyons, Text. Res. J. 34 (1964), p. 410.
[180] M.J. Schick, Text. Res. J. 44 (1974), p. 758.
[181] K.T. Chau, J. Eng. Mech. 123 (1997), p. 1.
[182] B.S. Gupta, *Frictional behavior of fibrous materials*, in *Polymer and Fibre Science: Recent Advances*, VCH Publishers, Inc., New York, NY, 1992, pp. 305–331.
[183] J. Skelton, Text. Res. J. 44 (1974), p. 746.
[184] A.E. Stearn, J. Text. Inst. 62 (1971), p. 353.
[185] C.M. Van Wyk, J. Text. Inst. 37 (1946), p. T285.
[186] O. Kallmes and H. Corte, Tappi. 43 (1960), p. 737.
[187] T. Komori and P.A. Makishima, Text. Res. J. 47 (1977), p. 13.
[188] G.M. Abbott and D. Nason, Text. Res. J. 56 (1986), p. 715.
[189] T. Komori and M. Itoh, Text. Res. J. 61 (1991), p. 420.
[190] B. Neckar, Text. Res. J. 67 (1997), p. 123.
[191] G.A. Carnaby and J.I. Curiskis, J. Text. Inst. 78 (1987), p. 293.
[192] G.H. Carnaby and N. Pan, Text. Res. J. 59 (1989), p. 285.
[193] J.W.S. Hearle and H.M.A.E. El-Behery, J. Text. Inst. 51 (1960), p. T164.
[194] J.I. Dunlop, J. Text. Inst. 65 (1974), p. 532.
[195] J.I. Dunlop, J. Text. Inst. 74 (1983), p. 92.
[196] M.D. Young, and A.D. Dircks, Text. Res. J. 55 (1985), p. 223.
[197] M.E. Yuksekkaya, *A novel technique for the assessing frictional properties of fibers*. Ph.D. thesis, North Carolina State University, Raleigh, USA, 1999.
[198] M.E. Yuksekkaya and W. Oxenham, *Analysis of mechanical and electrical noise interfacing the instrument during data acquisition: Development of a machine for assessing surface properties of fibres*, in *IEEE Conference*, IEEE Publishing, Atlanta, GA, May 1999.
[199] M.E. Yuksekkaya and W. Oxenham, *A novel technique for assessing the frictional characteristics of fibres. Part I: Development of instrument*, in *The 6th International Mechatronics Conference,* Middle East Technical University Publishing, Ankara, Turkey, Sep. 1999.
[200] M.E. Yuksekkaya and W. Oxenham, *A novel technique for assessing the frictional characteristics of fibres. Part II: Evaluation of the instrument*, The 6th International Mechatronics Conference, Middle East Technical University Publishing, Ankara, Turkey, Sep. 1999.
[201] M.E. Yuksekkaya and W. Oxenham, J. Text. Inst. 99 (2008) p. 545.
[202] M.E. Yuksekkaya, W. Oxenham, and M. Tercan, *Analysis of mechanical and electrical noise interfacing the instrument during data acquisition for measurement of surface properties of textile fibers*, Instrum. Measur. IEEE Trans. 57 no. 12 Dec. 2008, pp. 2885–2890.

AUTHOR SERVICES

Publish With Us

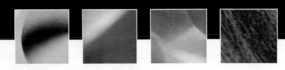

 Taylor & Francis
Taylor & Francis Group

 Routledge
Taylor & Francis Group

 Psychology Press
Taylor & Francis Group

informa
healthcare

The Taylor & Francis Group Author Services Department aims to enhance your publishing experience as a journal author and optimize the impact of your article in the global research community. Assistance and support is available, from preparing the submission of your article through to setting up citation alerts post-publication on **informa**world™, our online platform offering cross-searchable access to journal, book and database content.

Our Author Services Department can provide advice on how to:

- direct your submission to the correct journal
- prepare your manuscript according to the journal's requirements
- maximize your article's citations
- submit supplementary data for online publication
- submit your article online via Manuscript Central™
- apply for permission to reproduce images
- prepare your illustrations for print
- track the status of your manuscript through the production process
- return your corrections online
- purchase reprints through Rightslink™
- register for article citation alerts
- take advantage of our i*OpenAccess* option
- access your article online
- benefit from rapid online publication via i*First*

See further information at:
www.informaworld.com/authors

or contact:
Author Services Manager, Taylor & Francis, 4 Park Square, Milton Park, Abingdon, Oxon OX14 4RN, UK, email: authorqueries@tandf.co.uk

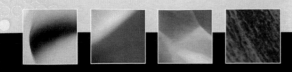

 Taylor & Francis
Taylor & Francis Group

International Journal of Fashion Design, Technology and Education

New journal in 2008

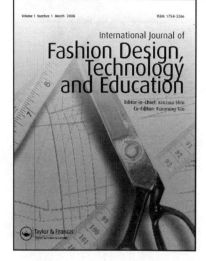

EDITOR:

Kristina Shin, *The Hong Kong Polytechnic University, China*

International Journal of Fashion Design, Technology and Education aims to provide a high quality peer-reviewed forum for research in fashion design, pattern cutting, apparel production, manufacturing technology and fashion education. The Journal will encourage interdisciplinary research and the development of an academic community which will share newly developed technology, theory and techniques in the fashion and textile industries, as well as promote the development of excellent education practice in the clothing and textile fields.

Contributions suitable for this new journal should fall into one of the following three categories:

- Research papers presenting important new findings
- Technical papers describing new developments or innovation
- Academic discussion papers dealing with medium to long-term trends and predictions.

To receive the table of contents for *International Journal of Fashion Design, Technology and Education* visit the journal homepage at www.tandf.co.uk/journals/tfdt

Submit your papers via Manuscript Central at http://mc.manuscriptcentral.com/tfdt

---

To sign up for tables of contents, new publications and citation alerting services visit **www.informaworld.com/alerting**

 **e**updates
Taylor & Francis Group

Register your email address at **www.tandf.co.uk/journals/eupdates.asp** to receive information on books, journals and other news within your areas of interest.

 Powered by
informaworld

For further information, please contact Customer Services at either of the following:
T&F Informa UK Ltd, Sheepen Place, Colchester, Essex, CO3 3LP, UK
Tel: +44 (0) 20 7017 5544 Fax: 44 (0) 20 7017 5198
Email: subscriptions@tandf.co.uk

Taylor & Francis Inc, 325 Chestnut Street, Philadelphia, PA 19106, USA
Tel: +1 800 354 1420 (toll-free calls from within the US)
or +1 215 625 8900 (calls from overseas) Fax: +1 215 625 2940
Email: customerservice@taylorandfrancis.com

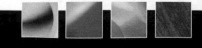

View an online sample issue at:
www.tandf.co.uk/journals/tfdt

Taylor & Francis
Taylor & Francis Group

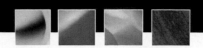

The Journal of the Textile Institute

Published on behalf of the Textile Institute

Increased to 8 issues per year

EDITOR-IN-CHIEF:

D. Buchanan, *North Carolina State University, USA*

The Journal of the Textile Institute welcomes papers concerning research and innovation, reflecting the professional interests of the Textile Institute in science, engineering, economics, management and design related to the textile industry and the use of fibres in consumer and engineering applications. Papers may encompass anything in the range of textile activities, from fibre production through textile processes and machines, to the design, marketing and use of products. Papers may also report fundamental theoretical or experimental investigations, practical or commercial industrial studies and may relate to technical, economic, aesthetic, social or historical aspects of textiles and the textile industry.

The Journal of
The Textile Institute
Editor David R. Buchanan 2007 Vol 98 No 1 ISSN 0040-5000

To sign up for tables of contents, new publications and citation alerting services visit **www.informaworld.com/alerting**

Register your email address at **www.tandf.co.uk/journals/eupdates.asp** to receive information on books, journals and other news within your areas of interest.

Powered by
informaworld

For further information, please contact Customer Services at either of the following:
T&F Informa UK Ltd, Sheepen Place, Colchester, Essex, CO3 3LP, UK
Tel: +44 (0) 20 7017 5544 Fax: 44 (0) 20 7017 5198
Email: **subscriptions@tandf.co.uk**

Taylor & Francis Inc, 325 Chestnut Street, Philadelphia, PA 19106, USA
Tel: **+1 800 354 1420 (toll-free calls from within the US)**
or **+1 215 625 8900 (calls from overseas) Fax: +1 215 625 2940**
Email:**customerservice@taylorandfrancis.com**

View an online sample issue at:
www.tandf.co.uk/journals/tjti